P.J.Lony

2424A
£28

PRECEDENCE IN ENGLAND AND WALES

PRECEDENCE IN ENGLAND AND WALES

G. D. SQUIBB

CLARENDON PRESS · OXFORD
1981

Oxford University Press, Walton Street, Oxford OX2 6DP
London   Glasgow   New York   Toronto
Delhi   Bombay   Calcutta   Madras   Karachi
Kuala Lumpur   Singapore   Hong Kong   Tokyo
Nairobi   Dar es Salaam   Cape Town
Melbourne   Auckland
and associate companies in
Beirut   Berlin   Ibadan   Mexico City

Published in the United States by
Oxford University Press, New York

© *G. D. Squibb 1981*

All rights reserved. No part of this publication may be reproduced, stored in a retrieval system, or transmitted, in any form or by an means, electronic, mechanical, photocopying, recording, or otherwise, without the prior permission of Oxford University Press

British Library Cataloguing Publication Data
Squibb, G.D.
  Precedence in England and Wales.
  1. Precedence - England
  I. Title
  929.7'2   CR3888
  ISBN 0-19-825389-3

Typeset by DMB (Typesetting)
Reproduced from copy supplied
Printed and bound in Great Britain
by Billing and Sons Limited
Guildford, London, Oxford, Worcester

To
The Most Noble
Miles Francis
Duke of Norfolk
C.B., C.B.E., M.C.
Earl Marshal and Hereditary Marshal of England
This Work is
By His Grace's Permission
Respectfully Dedicated
By His Lieutenant, Assessor, and Surrogate

Epigraph

Forsomuch as in all great Councils and congregations of men, having sundry degrees and offices in the Common wealth, it is very requisite and convenient that an order should be had and taken for the placing and sitting of such persons as are bound to resort unto the same to the extent that they knowing their places may use the same without displeasure or let of the Council.
Statute 31 Hen.VIII, c.10, preamble.

Precedency, like Forms in Parliament, considered only in itself, is ridiculous and vain: But considered as a Means to the Support of Dignity and Order, is essential to the very Existence of Authority. *The Question of the Precedency of the Peers of Ireland in England* (Dublin, 1739), p. 95.

Preface

This work is confined to precedence in England, because English law (which was applied to Wales by the Laws in Wales Act 1535) on this topic is different from that of Scotland. Not only would it be confusing to attempt to deal with the laws of both countries in one book, but an English lawyer must acknowledge himself to be unqualified to write on Scots law.

Coke and Hale wrote on the law relating to precedence as it was in the seventeenth century and Blackstone followed them in the eighteenth century. In 1851 Sir Charles Young, Garter King of Arms, had his *Order of Precedence with Authorities and Remarks* privately printed. Since Young's time there has been no further exposition of the authorities, the only guidance on the subject readily available being the tables of precedence appearing in periodical works of reference, which are not entirely consistent with each other. The present work began as an attempt to resolve the inconsistencies. This led to an examination of the manner in which the modern law of precedence has developed and the preparation of tables of precedence which differ in some respects from those set out in current works of reference. It is therefore hoped that what is primarily a work of legal history will also be of practical value to those who have to prepare lists of names or to arrange processions or functions at which proper precedence has to be observed.

Soon after publication table (ii) in Appendix IV will require amendment by the insertion of an entry for the Princess of Wales. Since there has never previously been a time when a Queen Mother or a Queen Dowager and a Princess of Wales have been contemporaries, there is no authority as to their relative precedence. The matter has therefore to be left until The Queen's pleasure has been declared.

Clause 13 of the Supreme Court Bill provides for the consolidation with amendments of the enactments referred to on pp. 55-6. Should it become law in its present form, the only substantive amendment will make it necessary to insert the Vice-Chancellor after the President of the Family Division in table (i) in Appendix IV.

I should like to place on record my gratitude to the Chapter of the College of Arms for generously giving me unrestricted access to the manuscript volumes known as the Earl Marshal's Books, in which are recorded the terms of the royal warrants which are the primary source of the modern law of precedence. I also have to thank two members of the College, Mr Colin Cole, Garter King of Arms, and Mr Sedley Andrus, Lancaster Herald, and Major-General Peter Gillett, Secretary of the Central Chancery of the Orders of Knighthood, Mr J. S. G. Simmons of All Souls College, and the late Dr F. W. Steer, Maltravers Herald Extraordinary, for help in a variety of ways, and Mrs Margaret Hogan for kindly devoting some of her well-earned retirement to preparing the typescript. Finally, I count it a privilege that my work has had the benefit of the care and skill of the staff of the Clarendon Press.

G.D.S.

Temple
1 June 1981

Contents

ABBREVIATIONS		xi
TABLE OF STATUTES AND MEASURES		xiii
TABLE OF CASES		xvii
INTRODUCTION		1
I.	GENERAL PRECEDENCE AMONG MEN BEFORE THE ACT OF SUPREMACY (1534)	7
II.	GENERAL PRECEDENCE AMONG MEN SINCE THE ACT OF SUPREMACY (1534)	23

    A. Personal
        (i) *The Sovereign and the Royal Family*   25
        (ii) *The Peerage*   30
        (iii) *The Baronetage*   36
        (iv) *The Knightage*   40
        (v) *Esquires*   42
    B. Official
        (i) *The Great Officers of State*   46
        (ii) *Diplomatic Representatives*   48
        (iii) *The Episcopate*   48
        (iv) *The Judiciary*   52
        (v) *Privy Councillors*   58
        (vi) *Other Office-Holders*   59

III.	GENERAL PRECEDENCE AMONG WOMEN	62
IV.	SPECIAL PRECEDENCE	72
V.	PERSONAL PRECEDENCE BY ROYAL WARRANT	81
VI.	TABLES OF PRECEDENCE	86
VII.	DISPUTES AS TO PRECEDENCE	90

APPENDICES
I.	THE LORD CHAMBERLAIN'S ORDER OF 1520, AS AMENDED IN 1595	98
II.	THE 'ANCIENTY' OF BARONIES BY WRIT OF SUMMONS	101

III: FORMS OF PETITIONS FOR PRECEDENCE
    A. Petitions for Royal Warrants of Precedence
        (i) *By the Brother and Sister of a Peer who has succeeded his Grandfather*   113
        (ii) *By the Brother and Sister of a Peer who has succeeded his Uncle*   114
        (iii) *By the Widow and Children of a Life Peer-designate*   115
        (iv) *By the Widow of the Heir Apparent of a Baronet*   116
        (v) *By the Widow of a Knight Bachelor-designate*   117
    B. Petition by a Baron by Writ for Precedence in Parliament   118

IV: MODERN TABLES OF PRECEDENCE
        (i) *Men*   119
        (ii) *Women*   123

BIBLIOGRAPHY   125

INDEX   131

Abbreviations

A.C. (preceded by date) Law Reports, Appeal Cases, 1891-
App.Cas.   Law Reports, Appeal Cases, 1875-90
Ashm.   Ashmolean
B.L.   British Library
Bl.Comm.   Blackstone's *Commentaries*
Bodl.   Bodleian Library
C.& P.   Carrington and Payne's Reports, 1823-41
C.P.   *Complete Peerage* (2nd edn.)
C.P.R.   *Calendar of Patent Rolls*
C.S.P.D.   *Calendar of State Papers, Domestic*
Cl. & F.   Clark and Finnelly's Reports, 1831-46
Co.Inst.   Coke's Institutes
Co.Litt.   Coke on Littleton
Co.Rep.   Coke's Reports, 1572-1616
Coll.Arm.   College of Arms
Collins   *Proceedings on Claims concerning Baronies by Writ*
E.& E.   Ellis and Ellis's Reports, 1858-61
E.E.T.S.   Early English Text Society
H.L.C.   House of Lords Cases, 1847-66
Harl.   Harleian
Hawk.P.C.   Hawkins's *Pleas of the Crown*
I.   I.Series in Coll.Arm.
I.T.   Inner Temple
Keb.   Keble's Reports, 1661-77
Keil.   Keilway's Reports, 1327-1578
L.J.   *Journals of the House of Lords*
Ld Raym.   Lord Raymond's Reports, 1694-1732
Lev.   Levinz's Reports, 1660-96
P. (preceded by date) Law Reports, Probate, Divorce and Admiralty Division, 1891-
Q.B. (preceded by date) Law Reports, Queen's Bench Division, 1891-
Rawl.   Rawlinson
Rolle   Rolle's Reports, 1614-25

Rot. Parl.	*Rotuli Parliamentorum*
S.R.&.O.	Statutory Rules and Orders
Stra.	Strange's Reports, 1716-47
Taunt.	Taunton's Reports, 1807-19
Wils.	G. Wilson's Reports, 1742-74
Y.B.	Year Book

TABLE OF STATUTES AND MEASURES

(The chapter numbers of the statutes before the reign of George I are those of *The Statutes of the Realm* (London, 1807) in all cases of statutes included in that edition.)

51 & 52 Hen.III	Dictum of Kenilworth (1266)	106
3 Edw.I, c.39	Statute of Westminster I (1275)	4
5 Ric.II, st.2, c.4	Parliament (1382)	12
11 Ric.II, c.1	Archbishop of York (1387)	14
12 Ric.II, c.2	Corrupt Appointments to Offices (1388)	19
16 Ric.II, c.4	Liveries (1392)	42
20 Ric.II, c.2	Liveries (1396)	42
1 Hen.VIII, c.14	Apparel (1509)	76
24 Hen.VIII, c.12	Ecclesiastical Appeals Act 1532	23, 25
	s.1	25
	s.3	25
26 Hen.VIII, c.1	Act of Supremacy (1534)	4, 23, 25
c.14	Suffragan Bishops Act 1534, s.2	56
27 Hen.VIII, c.26	Laws in Wales Act 1535	vii
31 Hen.VIII, c.10	House of Lords Precedence Act 1539	3, 23, 24, 26, 27, 31, 84, 90, 97, 103
	s.2	24, 27, 52
	s.3	48, 51
	s.4	27, 28, 29, 46, 47, 53, 60
	s.5	46, 60
	s.7	30, 101
	s.8	47
	s.10	24, 47, 60
1 Edw.VI, c.7	Justices of the Peace Act 1547, s.3	49
1 Mar.sess.3, c.1	Queen Regent's Prerogative Act 1554	62
1 & 2 Ph.& Mar., c.8	See of Rome (1554), s.12	25
1 Eliz.I, c.1	Act of Supremacy	23, 49
	s.1	25
	s.7	26
	s.9	26

5 Eliz.I, c.18	Lord Keeper Act 1562	53
3 Car.I, c.4 (private)	Earldom of Arundel (1627)	110
12 Car.II, c.9	Taxation (1660)	43
c.29	Taxation (1660)	43
14 Car.II, c.4	Act of Uniformity 1662, s.21	62
29 & 30 Car.II, c.15 (private)	Lord Audley (1677)	109
1 Will.& Mar., sess 1, c.21	Great Seal Act 1688, s.2	61
sess 1, c.3 (private)	Prince George of Denmark (1688)	26
sess 2, c.2	Bill of Rights (1688)	26
4 & 5 Ann., c.14	Princess Sophia (1705)	61
6 Ann., c.11	Union of Scotland Act 1706	39
	art.XXIII	31
c.16	Game (1706), s.4	43
10 Ann., c.8	Princess Sophia's Precedence Act 1711	6, 28, 70
1 Geo.I, c.3 (private)	Duke of Ancaster (1714)	46
39 & 40 Geo.III, c.67	Union with Ireland Act 1800	39
	art.IV	31, 49
44 Geo.III, c.54	Yeomanry (1804), s.36	43
53 Geo.III, c.24	Administration of Justice (1813), s.4	53
56 Geo.III, c.13	Naturalization of Prince Leopold (1816)	27
1 & 2 Will.IV, c.56	Bankruptcy Court (1831), s.2	54
5 & 6 Will.IV, c.76	Municipal Corporations Act 1835	
	s.57	76
	s.137	77
3 & 4 Vict., c.2	Naturalization of Prince Albert (1840)	27
4 & 5 Vict., c.52	Court of Chancery Act 1841, s.25	54
5 & 6 Vict., c.122	Bankruptcy (1842), s.65	54
8 & 9 Vict., c.100	Lunatics (1845)	57
9 & 10 Vict., c.95	County Courts Act 1846	56
10 & 11 Vict., c.102	Bankruptcy (1847), s.2	54
c.108	Ecclesiastical Commissioners Act 1847, s.2	48

14 & 15 Vict., c.4	Appointment of Vice-Chancellor (1851), s.2	54
15 & 16 Vict., c.89	Court of Chancery Act 1852	
	s.1	57
	s.53	54
17 & 18 Vict., c.100	Court of Chancery Act 1854	
	s.3	55
	s.5	55
18 & 19 Vict., c.13	Lunacy (1854), s.6	57
32 & 33 Vict., c.42	Irish Church Act 1869, s.13	50
36 & 37 Vict., c.66	Supreme Court of Judicature Act 1873	2
	s.5	55
	s.11	55
39 & 40 Vict., c.59	Appellate Jurisdiction Act 1876, s.6	32, 66
45 & 46 Vict., c.50	Municipal Corporations Act 1882	
	s.15 (5)	76
	s.163 (5)	76
	s.257 (2)	77
53 & 54 Vict., c.5	Lunacy Act 1890, s.111 (1)	57
4 & 5 Geo.V, c.59	Bankruptcy Act 1914, s.54 (1)	32
c.91	Welsh Church Act 1914	
	s.1 (1)	50
	s.2 (2)	50
6 & 7 Geo.V, (private; not printed)	Alexander's Restitution Act 1916	109
9 & 10 Geo.V, c.71	Sex Disqualification (Removal) Act 1919, s.1	70
12 & 13 Geo.V, c.60	Lunacy Act 1922, s.1 (1)	57
15 & 16 Geo.V, c.49	Supreme Court of Judicature (Consolidation) Act 1925	
	s.6 (2)	55
	s.16 (2)	55
	s.16 (2A)	55
	s.16 (3)	55
	s.16 (4)	56
	s.125	56
16 & 17 Geo.V, c.60	Legitimacy Act 1926	35
18 & 19 Geo.V, c.26	Administration of Justice Act 1928, s.14 (1)(e)	73

23 & 24 Geo.V, c.51	Local Government Act 1933	
	s.18 (5)	76
	s.302 (b)	77
2 & 3 Geo.VI, c.72	Landlord and Tenant (War Damage) Act 1939, s.4 (1)	32
11 & 12 Geo.VI, c.62	Statute Law Revision Act 1948	
	s.5	3
	Sch.2	3, 23
12, 13, & 14 Geo.VI, c.101	Justices of the Peace Act 1949	
	s.13 (3)	76
	s.13 (4)	76
14 & 15 Geo.VI, C.A.M.No.2	Bishops (Retirement) Measure 1951	52
6 & 7 Eliz.II, c.21	Life Peerages Act 1958	
	s.1 (1)	33
	s.1 (2)(a)	33
7 & 8 Eliz.II, c.73	Legitimacy Act 1959	35
	s.2	36
1963, c.48	Peerage Act 1963, s.3(1)(a)	32, 65
1970, c.31	Administration of Justice Act 1970	
	s.1 (1)	55
	s.5 (1)	56
	Sch.2	55
1971, c.23	Courts Act 1971	
	s.1	56
	s.1 (2)	75
	s.3	56, 78
	s.21	78
	Sch.2	56
1972, c.70	Local Government Act 1972	78
	s.3 (4)	76
	s.22 (4)	76
	s.219 (1)	75
	s.245 (1)	76
	s.245 (6)	76
1974, C.A.M. No.3	Church of England (Worship and Doctrine) Measure 1974, s.1 (7)	62
1978, C.A.M. No.1	Dioceses Measure 1978, s.15 (2)(a)	52

TABLE OF CASES

Anon. (1611), 12 Co.Rep. 81	37
——— (1742), 2 Stra.176	74
Ashton v. Jennings (1674), 3 Keb.462;	
2 Lev.133	1, 45, 62, 66, 90, 91
Att.Gen. v. Lord Advocate (1834), 2 Cl. & F.481	3, 74
Beauchamp (Barony), [1925] A.C. 153	105, 106, 107
Beaumont Peerage (1840), 6 Cl. & F.268	109
Birmingham (Mayor), Ex parte (1860) 3 E. & E.222	76
Blackburn v. Flavelle (1881), 6 App.Cas. 628	39
Botreaux, Hungerford, etc.	
Peerage Case (1871)	107, 108, 109, 110, 111, 112
Buckingham's (Duke of) Case (1514), Keil.170	90
Burgh, Strabolgi and Cobham Baronies (1913)	102, 104, 109, 112
Camoys Peerage (1839), 6 Cl. & F.769	107, 110
Caudrey's Case (1591), 5 Co.Rep.1	23
Chune v. Pyot (1615), 1 Rolle 237	75
Clifton Peerage Case (1673), Collins 291	102
Cowley (Earl) v. Cowley (Countess), [1901] A.C.450	64
Cromwell (Barony) Peerage Claim (1922)	110
Devon Peerage Case (1831), 2 Dow. & Cl.200	106
Dudley Peerage Claim (1914)	111
Fauconberg, Darcy (de Knayth), and Meinell Baronies (1903)	110, 111
Fitzwalter Peerage Case (1844), 10 Cl. & F.946	108
Furnivall (Barony) (1912)	108
Gas Float Whitton No.2, [1898] P.42	2
Grey de Ruthyn Peerage Case (1876)	110
Hastings Peerage Case (1841), 1 Cl. & F.144	105, 106, 108
Latymer Peerage Claim (1912)	111
Mackonochie v. Penzance (Lord) (1881), 6 App.Cas.424	4
Manchester Corporation v. Manchester Palace of Varieties Ltd, [1955] P.133	2, 91
Messor v. Molyneux (1741)	43
Montacute and Monthermer Peerages Case (1928)	104
Mowbray and Segrave Peerage Claims (1877)	106, 108, 112

xviii   *Table of Cases*

Nevil's Case (1604), 7 Co.Rep. 33a	33
Paston v. Ledham (1459), Y.B. 37 Hen.VI, Pasch., pl.8	1
Phillips v. Halliday, [1891] A.C.228	4
Poole and Redhead's Case (1614), 1 Rolle 87	90, 91, 92
Precedence of Serjeants (1840), 9 C.& P. 371 n	93
Proclamations (Case of) (1611), 12 Co.Rep. 74	3
Prohibitions del Roy (1607), 12 Co.Rep. 63	94
R. v. Brough (1748), 1 Wils. 244	43
— v. Comptroller-General of Patents, [1899] 1 Q.B.909	73
— v. Knollys (1694), 1 Ld Raym.10	3, 31
— v. Parker (1668), 2 Keb.316	1
Roos (Barony) (1616), Collins 162	103, 105
Rutland's (Countess of) Case (1605), 6 Co.Rep. 52b	64
St.John Peerage Claim, [1915] A.C.282	105, 107
Shrewsbury Peerage Case (1857)	82
Strabolgi Barony (1915)	110
Strange of Knokin and Stanley (Baronies) (1920)	104, 109, 111
Talbot v. Eagle (1809), 1 Taunt. 510	43
Vaux Peerage (1837), 3 Cl. & F.526	107, 112
Wensleydale Peerage Case (1856), 5 H.L.C. 958	3
Wiltes Claim of Peerage (1869), L.R.4 H.L. 126	106

Introduction

On 10 May 1674 Rainsford and Wilde JJ., sitting in the Court of King's Bench, found themselves faced with a case the like of which seems never before to have been the subject of litigation in a court of common law. Mrs Margaret Ashton, the wife of Roger Ashton, a doctor of divinity of Cambridge University and vicar of St. Andrew's, Plymouth, had been invited to a funeral. There she met Mrs Jennings, the wife of a justice of the peace. They disagreed about who should have the higher place. Mrs Ashton refused to give way, so Mrs Jennings '*molliter manus imposuit*' to put Mrs Ashton out of her place. This was not to be borne by the Ashtons, so Dr Ashton brought an action against Mr Jennings for the battery. It was argued for the defendant that the wife of an esquire and a justice of the peace ought to have precedence and take place of the wife of a doctor of divinity, for though a doctor of divinity took place of an esquire, that was in respect of his degree, which was personal to himself only and not communicable to his wife. The court gave judgment for the plaintiff, because the defendant had confessed the battery by his plea, and avoided deciding the delicate question of precedence by holding that it was a matter within the jurisdiction of the 'Court of Honour'.[1]

The decision in *Ashton* v. *Jennings*, although it left unresolved the real point in issue between the parties, is important in two respects. The first is that it was never even suggested that the question of the relative precedence of Mrs Ashton and Mrs Jennings was not a matter of law. Indeed, that the court regarded it as a matter of law is implicit in its statement that the matter fell within the jurisdiction of the Court of Chivalry, for as Keeling C. J. had said seven years before, 'This Court of honor is part of the law of England',[2] a statement described by Sir George Clark as 'a solid

---

[1] *Ashton* v. *Jennings* (1674), 3 Keb.462; 2 Lev.133. Levinz wrongly assigned the case to Trinity Term 1675, but in other respects his is the better report. For Ashton, see A. G. Matthews, *Walker Revised* (Oxford, 1948), p. 77.

[2] *R.* v. *Parker* (1688), 2 Keb.316. Cf. *Paston* v. *Ledham* (1459), Y.B. 37 Hen.VI, Pasch., pl.8, where Nedham J. said: 'The law of the Constable and Marshal is the law of the land and the law of our Lord the King.... We take notice of it.'

constitutional fact'.³ The other important feature of the decision in *Ashton* v. *Jennings* is that the court, by holding that matters of precedence were within the jurisdiction of the Court of Chivalry, also held by necessary implication that the law relating to precedence is not common law, but civil law, by which proceedings in the Court of Chivalry are regulated.⁴

As will appear later, there was no historical justification for holding that the Court of Chivalry had jurisdiction in matters of precedence.⁵ Nevertheless, it has been decided that the Court of Chivalry has such jurisdiction, and the decision has never been overruled, so it is necessary briefly to consider its effect.

While the Court of Chivalry is a civil-law court in that its procedure is that of the civil law, the substantive law which it administers can only be regarded as civil law in the somewhat artificial sense that modern admiralty law can still be regarded as civil law, even though it is to be ascertained from the practice and judgments of its judges⁶ and notwithstanding the transfer of the jurisdiction of the Court of Admiralty to the High Court of Justice by the Supreme Court of Judicature Act 1873. That the jurisdiction of the Court of Chivalry was not so transferred at the same time was no doubt due to the belief, held in *Manchester Corporation* v. *Manchester Palace of Varities Ltd.*⁷ to have been mistaken, that the Court was obsolete. Even if the jurisdiction had been transferred by the Act of 1873, the substantive law would still not have been common law, but civil law. Thus it can almost be said that as a result of the decision of the Court of King's Bench in 1674 the substantive law of precedence in England is masquerading as civil law.

There are passages in the *Corpus Juris Civilis* which deal, either explicitly or implicitly, with matters of precedence, and civilians have endeavoured to construct from them a coherent statement of law on the subject.⁸ There is, however, but little practical assistance to be derived from such academic essays. The substantive

---

³ Sir G. Clark, *Three Aspects of Stuart England* (London, 1960), p. 31.
⁴ A. Duck, *De Usu et Authoritate Juris Civilis* (Leyden, 1654), p. 395 and the authorities there cited. The authorities are omitted from the translation of the part of Duck's work relating to England appended to J. Beaver, *History of the Roman or Civil Law* (London, 1724).
⁵ See pp. 90-93 *post*.
⁶ *The Gas Float Whitton No. 2*, [1898] P.42, per Lord Esher, M.R., at p. 48.
⁷ [1955] p. 133.
⁸ e.g. R. Zouche, *Elementa Jurisprudentiae* (Amsterdam, 1652), p. 342; *Jurisprudentia Heroica* (Brussels, 1668), pp. 50b, 51, 422v., *et seq*.

law of precedence in England is not, and never has been, the law of ancient Rome. Like the law relating to heraldic matters, the substantive law is based on English practice. As in the case of heraldic law, it can only be regarded as civil law by the application of the rule laid down by the medieval civilians that matters of honour and dignity were to be ordered and ruled according to the custom of every particular country.[9]

There is no historical basis for so regarding the law relating to precedence as civil law. To do so would be to ignore the fact that in medieval England there was no one lawful authority with regard to precedence and that neither derived its power from the civil law.

As far as lay people, whether men or women, were concerned, the fountain of honour was the King. In the words of the preamble to the House of Lords Precedence Act 1539,[10] it appertained to the 'prerogative Royall, to give such honour, reputation, and placing to his Counsellers, and other his Subjects, as shall be seeming to his most excellent wisedome'. It was held in *R.* v. *Knollys*[11] that the Crown had this power by the common law, but was bound by the Act of 1539 in the exercise of the power. Furthermore, the power is limited by the common law itself.[12] Thus Queen Victoria was held to have exceeded her power when she purported to create a life peerage with a seat in Parliament.[13]

The basis of the law relating to the precedence of lay people is the usage observed at the courts of the Norman kings, which, according to modern theory, must be regarded as evidence of grants by those kings in the exercise of the royal prerogative. As Lord Brougham L.C. said in *Att.-Gen.* v. *Lord Advocate*,[14] immemorial usage with regard to precedence is equivalent to a grant from the Crown, the fountain of all honour. Evidence of the observance of such usage from time immemorial in the legal sense,

---

[9] A. Collins, *Proceedings ... on Claims ... concerning Baronies by Writ* (London, 1734), p. 63. See also G. D. Squibb, *High Court of Chivalry* (Oxford, 1959), pp. 163-5.
[10] 31 Hen.VIII, c.10. The short title was given by the Statute Law Revision Act 1948, s.5, Sch.2.
[11] (1694), 1 Ld Raym. 10, at p. 16.
[12] 'The King hath no prerogative but that which the law of the land allows him': *Case of Proclamations* (1611), 12 Co.Rep.74.
[13] *Wensleydale Peerage Case* (1856), 5 H.L.C. 958. The grantee, Sir James Parke, was later created Baron Wensleydale of Walton with the usual remainder (*C.P.* xii, pt.ii.496).
[14] (1834), 2 Cl.& F. 481, at p. 484.

i.e. from before 1189, the first year of the reign of Richard I,[15] is scanty. Indeed, the usage with regard to the precedence of dukes, marquesses, and viscounts and members of their families is manifestly later, since those ranks in the peerage did not exist in 1189. Nevertheless, usage which must have originated after 1189 can by analogy with the law relating to prescription be regarded as evidence of a lost grant made after that date, for there is a presumption of law that where long enjoyment of an alleged right can be shown it must have had a legal origin, if such an origin is possible.[16]

Such is the manner in which post-medieval common lawyers have expounded the law. It seems, however, more likely that the early law relating to precedence was, like the forest law, not common law, but merely the expression of the king's will. As Richard fitz Nigel, writing in the last quarter of the twelfth century in his *Dialogus de Scaccario*, said: 'The forest has its own laws, based, it is said, not on the Common Law of the realm (*non communi regni iure*), but on the arbitary legislation of the King; so that what is done in accordance with forest law is not called "just" without qualification, but just according to forest law.'[17] This may well be the true explanation of why the common-law judges have always declined jurisdiction with regard to non-statutory precedence.

The basis of the precedence of ecclesiastical dignitaries was entirely different. Their precedence was governed by canon law until the breach with Rome, when in 1534 Henry VIII became the statutory 'Supreme Head in earth of the Church of England, called *Anglicana Ecclesia*',[18] and the rules of the canon law, in so far as they were left unaltered by Parliament and were allowed by general consent and custom within the realm, became part of the King's ecclesiastical law.[19]

When the scale of ecclesiastical precedence was a matter of canon law, it necessarily remained distinct from that relating to laymen. Once the King alone became the sole source of precedence, it became possible to have a single scale applicable to

---

[15] This date was established by analogy with the period of limitation, fixed by the Statute of Westminster I (1275), for the bringing of writs of right.
[16] *Phillips* v. *Halliday*, [1891] A.C. 228, per Lord Herschell at p. 231.
[17] Translated in C. Johnson (ed.), *Dialogus de Scaccario* (London, 1950), pp. 59-60.
[18] 26 Hen. VIII, c.1.
[19] *Mackonochie* v. *Penzance (Lord)* (1881), 6 App.Cas.424, per Lord Blackburn at p. 448.

both churchmen and laymen.[20] Hence what may appear at first sight to be a strange chronological division at the Reformation of the first two chapters of this book. In this context the relevance of the Reformation is not theological, but juridical.

Although it is probably but a coincidence, there is a change during the Reformation period in the sources from which the substantive law of precedence is to be ascertained. Until the time of Henry VIII there is no legislation relating to precedence. The law has to be ascertained from evidence of usage and from statements of what the law was believed to be at various times, though such statements were not treatises on the subject by lawyers, but guides or sometimes directions to those whose duty it was to arrange ceremonies and processions. These statements are not always consistent with each other. The variations led George Edward Adams (later Cokayne), when Rouge Dragon Pursuivant (1859-69), to believe that the order of precedence had varied from time to time, [21] but it seems more likely that the variations are due to divergent views on doubtful points rather than to intentional changes initiated by the Crown.

In addition to explicit statements by contemporaries, some assistance can also be obtained from accounts of processions and instructions for the marshalling of processions, in which names are arranged in order of rank, but these have to be used with caution, since a procession is sometimes marshalled in a manner special to its occasion.[22]

From the reign of Henry VIII there is legislation by which the general order of precedence as it had become settled on the eve of the Reformation has been considerably elaborated to provide for the new ranks and offices which have been created during the following centuries. This has been done sometimes by royal commands, usually expressed in the form of warrants under the

---

[20] It was, of course, unnecessary to devise a single scale for women, since there were no longer any women with ecclesiastical precedence after the dissolution of the nunneries.
[21] 38 E.E.T.S. (1868), p. 236.
[22] See p. 16 *post*. Since processions are usually marshalled in reverse order of precedence, in ascending order, with the lower ranks before the higher, in order to avoid confusion it is well not to use the words 'before' and 'after' when referring to a processional as evidence of relative precedence, but to convert the processional notionally into a table in the usual descending order by using the words 'above' and 'below'. The words 'before' and 'after' are, however, always used in official documents in respect of higher and lower precedence.

6 Introduction

sign manual, sometimes by Acts of Parliament, and sometimes by Orders in Council.[23]

The nature of all these changes brought about by the Crown alone and by the Crown and the two Houses of Parliament together is essentially the same. They are supplementary to the basic rules, defining the precedence which they confer by reference to precedence already existing. There has never been any comprehensive legislation on the subject in England comparable to the royal warrant of 9 March 1905, which settled precedence in Scotland. The English legislation has merely placed new categories of persons among those already entitled to precedence. A good example of this procedure is to be seen in the statute of 1711, by which the Electress Sophia of Hanover, her son George (afterwards King George I), her grandson George (afterwards King George II), and the heirs of her body, being protestants, were given rank and precedence after the issue of Queen Anne and before the Archbishop of Canterbury, the Great Officers of State, and the dukes and other peers.[24]

Not all the changes which have occurred in the law of precedence have been brought about by legislation. Some ranks, e.g. knights bannerets, and some offices, e.g. that of the Master of the Court of Wards and Liveries, to which precedence was attached have ceased to exist, so that the modern law has to be ascertained by taking the medieval law as a basis and proceeding from it by addition and subtraction.[25]

---

[23] For the procedure on the grant of warrants under the sign manual, see pp.   *post*.
[24] Princess Sophia's Precedence Act 1711.
[25] The tables of precedence in Appendix IV have been so constructed.

CHAPTER I

PRECEDENCE AMONG MEN BEFORE THE ACT OF SUPREMACY (1534)

That precedence was a matter of law in Anglo-Saxon England is shown by a document entitled 'Of People's Ranks and Law' drawn up at some time between about 1029 and 1060. It is there stated:'It was whilom, in the laws of the English, that people and law went by ranks, and then were the counsellors of the nation of worship worthy, each according to his condition, eorl and ceorl, thegen and theoden.'[1]

The order of ranking before the Norman Conquest can be observed in the witness lists of royal charters. While no one charter discloses the complete order, a number of charters when read together show that the order after the King was members of the royal family, archbishops, bishops, abbots, ealdormen, and the King's thegns. Although it is not without interest to observe that the ecclesiastics appear in a separate group after members of the royal family and before all other laymen, a feature which is found after the Norman Conquest, an investigation of the Anglo-Saxon law of precedence is neither necessary nor helpful for a proper understanding of the modern law on the subject, for there is nothing in the modern law which has an Anglo-Saxon origin. Precedence among ecclesiastics was the same throughout Western Europe. So far as laymen were concerned, precedence is a matter in which a fresh start was made at the Norman Conquest. Thereafter precedence was determined by the King as the fountain of honour.[2]

Precedence at the court of the Conqueror in England would not have differed from that at his court in Normandy. Although William I claimed to be king of England as the legitimate heir

---

[1] W. Stubbs, *Select Charters* (9th edn., Oxford, 1913), p. 88, where the Anglo-Saxon text is also printed. A similar document is printed in D. Whitelock, *English Historical Documents c. 500-1042* (2nd edn., London, 1979), pp. 468-71.
[2] 4 Co.Inst.381.

of Edward the Confessor, not only were his courtiers in Normandy also his courtiers in England, but the clerks who drafted his documents relating to English affairs immediately after the Conquest are also likely to have been newcomers. That a few of the Confessor's nobles, such as Edwin of Mercia, Morcar of Northumbria, and Waltheof of Northampton, came to terms with the Conqueror would not have had any influence on the law or custom observed at his court in England, if indeed that court can properly be regarded as distinct from his court in Normandy.

Precedence is not a subject on which the earliest post-Conquest legal authorities have anything to say. Nor is it a subject the existence of which is specifically noticed in contemporary non-legal sources. The way in which attention was paid to precedence in England before the end of the fourteenth century can only be a matter of inference from the way in which lists of names in documents are ordered.

The earliest post-Conquest lists of names which indicate the precedence observed in England are those of the signatories of and witnesses to diplomas, charters, and writs. Naturally, not all ranks are represented in every document, but the order in which they appear is invariable—the King, the Queen, the King's sons, archbishops, bishops, abbots, other ecclesiastics, earls, barons, and other laymen.

There is thus little to be derived from these early documents beyond the fact that precedence was regarded as a matter of moment by the Anglo-Norman kings. There does not appear to be any direct mention of the subject of precedence in England earlier than the statement at the end of the twelfth century by Richard fitz Nigel, Bishop of London in his *Dialogus de Scaccario* that in Domesday Book (compiled in 1086) the King's name heads the list, followed by those of the nobles who hold of him in chief, according to their order of dignity (*secundum status sui dignitatem*).[3] Since the list at the beginning of the entries for each county was copied from the circuit return,[4] the making of such lists must have been included in the instructions given to the commissioners for each of the circuits into which the country

---

[3] C. Johnson (ed.), *Dialogus de Scaccario*, (London, 1950). pp. 63-4.
[4] V. H. Galbraith, *The Making of Domesday Book* (Oxford, 1961), pp. 193, 195.

was divided. This implies an expectation that the commissioners or their clerks would have some knowledge of the order of dignity of the tenants-in-chief.

The Domesday lists of tenants-in-chief are drawn up according to the method used by the clerks in William I's chancery. After the King, the lists set out the ecclesiastics from the archbishops downwards, followed by the laymen, also in descending order, with women after men in their respective categories.[5]

From about 1100 there is another indication of the precedence of classes of persons. This is in the salutations of royal charters, the usual form (in translation) being: 'The King to all archbishops, bishops, abbots, earls, barons, justices, sheriffs, and other our faithful subjects, Greeting'. Sometimes the list is extended to include priors after abbots, and reeves and bailiffs after sheriffs. This grouping of ecclesiastics and laymen does not, of course, indicate that all ecclesiastics took precedence over all laymen. The true position was that there were two orders of precedence in England, one ecclesiastical and the other lay, the former being universal and the latter local. Although the precedence of laymen was but local, it would be anachronistic to describe it at this period as English: it should properly be regarded as Anglo-Norman. The combination of the two into one English order was still far in the future.

The setting out of ecclesiastics and laymen in separate orders of precedence in the salutations of charters continued for many centuries, with the addition of dukes in the reign of Edward III, marquesses in that of Richard II, and viscounts in that of Henry VI. This practice was followed by the clerks who prepared the earliest journals of the House of Lords, where the names of the lords spiritual present appear in a column on the left of the page and those of the lords temporal in the right. This arrangement is said to reproduce the seating arrangements as seen from the bar of the House with the lords spiritual on the King's right and the lords temporal on his left.[6]

Although the two archbishops had their place at the head of the ecclesiastical hierarchy in England by virtue of the universal

---

[5] Exceptionally, in the Nottinghamshire list in Great Domesday and in the Norfolk and Suffolk lists in Little Domesday, the earls precede the bishops (Galbraith, op.cit., p. 195).

[6] J. E. Powell and K. Wallis, *House of Lords in the Middle Ages* (London, 1968), p. 545.

law of the Church, their precedence between themselves was not so governed. This lacuna gave rise to a complicated dispute between Archbishop Lanfranc of Canterbury and Archbishop Thomas of York as to the primacy. This dispute was settled in 1072 by a written submission made by Thomas, in which he made an absolute profession of obedience to Lanfranc and his successors. Despite his submission, Thomas refused in 1093 to recognize Lanfranc's successor, Anselm, as primate. After its resurrection the dispute dragged on until 1126, when it was decided in favour of York, for the evidence on which Canterbury relied was forged.[7] Canterbury could, however, in his capacity of papal legate still claim the obedience of York. The quarrel about the relative precedence of the archbishops was so bitter at the Council of London in 1176 that their retainers came to blows, and the Council was broken up.[8] In 1324 Edward II ordered the cross of Canterbury to be borne before that of York.[9] The matter was finally decided in favour of Canterbury by a composition in 1353.[10]

So far as bishops were concerned, it had been decreed at the second Council of Milevis in 416, the second Council of Braga in 563, and the fourth Council of Toledo in 633 that at such assemblies bishops should take precedence according to the dates of their consecrations, except where special preference was due on account of the ancient customs or special privileges of their churches.[11] Such was the general rule. The question of the precedence of the English bishops was the first item on the agenda for the Council of London in 1075. It was then decided that the Archbishops of Canterbury and York were to be flanked by the Bishop of London on the left and the Bishop of Winchester on the right, and that if the Archbishop of York should be absent, the Bishops of London and Winchester would sit on the right and left respectively of the Archbishop of Canterbury.[12]

The only bishops at this period and for long afterwards were

---

[7] For accounts of the controversy and of the forgeries, see A. J. Macdonald, *Lanfranc* (Oxford, 1926), pp. 70-94, 271-91. It is but fair to the memory of Lanfranc to say that it is unlikely that he was responsible for forged evidence: see Z. N. Brooke, *The English Church and the Papacy* (Cambridge, 1931), pp. 123-5.

[8] William of Newburgh, *Historia Rerum Anglicarum* (82 Rolls Series, 1884), pp. 203-4.

[9] *Rot. Parl.* i.418a.

[10] D. Wilkins, *Concilia Magnae Britanniae* (London, 1737), iii.31.

[11] Macdonald, op.cit., pp. 98-9.

[12] William of Malmesbury, *De Gestis Pontificum Anglorum* (52 Rolls Series, 1870), p. 67.

diocesans, so that the precedence laid down in 1075 was all-embracing. It was not until towards the close of the thirteenth century that the Popes started the practice of consecrating bishops *in partibus infidelium*, assigning to them the titles of former sees which had ceased to be effective through being overrun by the pagans or for some other reason. These titular bishops were appointed to assist diocesan bishops. They first appeared in England in 1259, when Augustine, Bishop of Laodicea was appointed to officiate in the diocese of Durham, and they continued to be appointed until the Reformation. Such a titular bishop was sometimes described as the suffragan of the bishop in whose diocese he officiated, but this description, although technically correct, can be confusing, because the word 'suffragan' is also applicable to a diocesan bishop in his relationship with his metropolitan. Some English bishops were also assisted by bishops of Irish dioceses. Although Ireland could not exactly be described as *pars infidelium*, these Irish bishops were unable to obtain effective access to their sees and ranked in England with bishops *in partibus*. Both types of titular bishops had precedence after diocesan bishops.[13] If an English bishop became a cardinal, he did not thereby take any precedence in Parliament, but continued to rank there in right of his bishopric.[14]

When the Order of the Garter was instituted by Edward III in 1348 it was provided that the Bishop of Winchester should be the Prelate of the Order, and that he should have precedence after the archbishops and before all other bishops.[15] Notwithstanding this royal edict, the Bishop of London continued to claim the place which he had been given in 1075, as is witnessed by a note in the roll of the names and arms of the lords spiritual and temporal in the Parliament of 1515 that 'The bishopp of London claymyth to have preemynence in sittyng before all other bishoppys of the provynce of Cant' as cancellarius episcoporum.'[16] This claim may have been founded on a contention that the grant of precedence to

---

[13] A. Hamilton Thompson, *English Clergy in the Later Middle Ages* (Oxford, 1947), p. 49; *Catholic Encyclopedia* (London, [1911]), xii. 371-2; *New Catholic Encyclopaedia* (New York, etc., 1967), xi. 703-4.
[14] 4 Co.Inst. 362.
[15] Constitutions relating to the Officers of the Order, c.1, printed in E. Ashmole, *Institution, Laws and Ceremonies of the Most Noble Order of the Garter* (London, 1672), Appendix.
[16] Coll.Arm. MS.Box 40, no.42, reproduced in T. Willement, *Fac Simile of a Contemporary Roll ... of the Spiritual and Temporal Peers* (London, 1829), unpaginated.

the Bishop of Winchester in the Garter 'Constitutions' was uncanonical as being a usurpation of the papal authority over episcopal precedence. Whatever may have been the foundation of the claim, it was accepted in the parliament of 1523, the roll for that year placing London after Canterbury and before Winchester,[17] but the matter was not finally resolved until after the Reformation.[18]

Bishops were followed by abbots and priors.[19] The precedence of abbots among themselves depended on the dates of the foundation of their respective monasteries.[20] Presumably this rule also applied to priors.

The early precedence of earls and barons before all other laymen continued until 1337, when Edward III created his eldest son, Edward, the Black Prince, Duke of Cornwall. The creation charter contained no mention of precedence, for no such mention was required, since the Prince as the King's son already had precedence before all earls. This charter cannot, therefore, be regarded as introducing a new step in the scale of precedence. The first English duke who did not already have higher precedence as a member of the royal family was Thomas, Lord Mowbray, who was created Duke of Norfolk in 1397. The precedence of dukes as such had, however, already been recognised by 1382, when the statute 5 Ric.II, st.2, c.4 provided for the amercement of any archbishop, bishop, abbot, prior, duke, earl, or baron who did not obey his writ of summons to Parliament.

Although the early documents indicate the relative precedence of certain classes of persons, they throw little, if any, light on the precedence of persons within their respective classes. This is demonstrated by two diplomas issued early in the reign of William I. The first of these was issued at Westminster on Whit Sunday 1068 to the London church of St. Martin-le-Grand.[21] The episcopal signatories of this diploma were William of London, Odo of Bayeux, Hugh of Lisieux, Gosfrid of Coutances, Herman of Sherborne, Leofric of Exeter, and Giso of Wells, in that order. Had the English bishops in the group been arranged in the order

---

[17] Coll.Arm.MS. Box 40, no.41, reproduced in Powell and Wallis, *House of Lords in the Middle Ages*, pl.xix.
[18] See p. 48 *post*.   [19] e.g. Statute 5 Ric.II, st.2, c.4.
[20] For the dispute as to precedence between the abbots of Glastonbury and Westminster, see J. Armitage Robinson, *Somerset Historical Essays* (London, 1921), pp. 1 *et sqq*.
[21] *Regesta Regum Anglo-Normannorum* (Oxford, 1913), i.6.

General Precedence among Men before 1534 13

of their consecrations, which later became the usual order, the ranking would have been Herman of Sherborne (1045), Leofric of Exeter (1046), William of London (1051), and Giso of Wells (1061). There seems to be no apparent reason for ranking these bishops in the order in which they appear in the diploma.

The second diploma was granted to Bishop Giso of Wells at Whitsuntide 1068. Here the names of the bishops appeared in the following order—Odo of Bayeux, Gosfrid of Coutances, Herman of Sherborne, Leofric of Exeter, Ægelmar of Elmham, William of London, and Remigius of Dorchester.[22] With these diplomas may be compared a grant in diploma form made to Bishop Leofric of Exeter in 1069, in which the names of the bishops appear in yet another order—Odo of Bayeux, Herman of Sherborne, Leofric of Exeter, Gosfrid of Coutances, Giso of Wells, and William of London.[23]

It is clear that the scribes who drew these documents paid little or no attention to the relative seniority of the bishops who witnessed them. Even the first place occupied by Odo in all three is shown to have been accidental by another diploma of 1069 in which Odo appears after his brethren of London, Sherborne, Wells, and Exeter.[24]

The ranking of the earls (or counts) in these documents is equally inconsistent. In the first they are Hereford, Mortain, Mercia, Eu, Northumbria, Northampton, Montgomery, and Shrewsbury; in the second they are Hereford, Northampton, Mercia, Mortain, and Montgomery; and in the third Mortain, Hereford, Mercia, and Northampton. Such lists therefore throw no light on the relative precedence of earls, and they are presumably equally uninformative with regard to the precedence of barons among themselves, but so far as barons are concerned there is no means of checking this. When barons had become the lowest rank in the peerage their precedence was governed by the order of their creation, but barons during the reigns of the Norman kings and for long afterwards were but the ill-defined class of the greater tenants-in-chief, of whose relative precedence there seems to be no surviving evidence.

Similarly the lists of tenants-in-chief in Domesday afford no evidence as to the precedence of persons within their respective

[22] Ibid. i.7.   [23] Ibid. i.8.   [24] Calendared ibid.

classes. Thus, in the Dorset list the earls and counts appear in the order Alan (of Brittany), (Robert of) Mortain, and Hugh (of Chester), while in the Wiltshire list they appear in the order Mortain, Roger (of Montgomery), Hugh, and Alan. There is a similar lack of consistency in other counties.

As time went on, the draftsmen of formal documents came to take pains to get the names of persons mentioned in them in the correct order of personal precedence. That this was being done by the end of the fourteenth century is indicated by the citation by counsel on behalf of the Earl of Warwick in his dispute with the Earl Marshal in 1425 of letters patent granted by Richard II to Thomas, Duke of Gloucester, and a letter to the Pope of the same period. Counsel also relied on the order in which the lords appellants were named in the statute 11 Ric.II, c.1.[25]

Such meagre information regarding precedence as is to be gleaned from the lists of tenants-in-chief in Domesday Book and other early documents is purely accidental. It is not until the end of the fourteenth century that we come to a document dealing specifically with precedence. This is 'The Order of all Estates of Nobles and Gentry of England', dated 'the viijth of October in the year of our Lord God 1399'.[26] The original of this document was said to have been in the custody of Sir Richard St George, Norroy King of Arms from 1604 to 1623.[27] It is now known only from copies.[28] Some copies have the regnal year 23 Richard II added to the date. This is self-contradictory, for there was no 8 October in the twenty-third year of the reign of Richard II: his twenty-third year began on 22 June 1399 and was ended by his deposition on the following 29 September. If the 'Order' is correctly dated 8 October 1399, it may have been prepared for some purpose connected with the coronation of Henry IV, which took

---

[25] *Rot.Parl.* iv. 267.   [26] I.T. MS. Petyt 538/44, fo.29.

[27] Ibid., fo.71v. The statement in J. Conway Davies, *Catalogue of Manuscrips in the Library of ... the Inner Temple* (Oxford, 1972), ii.840 that the original was in the hands of Sir Henry St. George is incorrect. Sir Henry St. George held another document referred to on fo.29v. of the manuscript.

[28] e.g. in addition to the Inner Temple manuscript, Bodl.MS.Ashm.857, F.137, and Coll.Arm. MS. Vincent 151, p. 121, the latter printed in [C. G. Young], *Ancient Tables of Precedency* (n.p., n.d.), pp. 3,4. Young also printed the first parts of this and the other tables hereafter referred to in his *Privy Councillors and their Precedence* (n.p., 1860), pp. 41-7, but omitted the second parts relating to women. For the precedence of women, see pp. 62-71 *post*.

place six days later. This 'Order' is noteworthy because it is apparently the earliest evidence of the precedence enjoyed by the sons of peers and by persons below the rank of baron.

The 'Order' may have been derived from some earlier document, but since it includes 'My Lord Marquesse: a new Honor', it cannot have had an exact prototype earlier than 1 December 1385, when the first English marquessate was conferred upon Robert de Vere, Earl of Oxford, who was created Marquess of Dublin. There was no marquess between 13 October 1386, when de Vere was promoted to be Duke of Ireland, and 29 September 1397, when John Beaufort, Earl of Somerset was created Marquess of Dorset. Beaufort was the only marquess living on 8 October 1399. His eldest son was not born until 1401,[29] so the entries for the eldest and younger sons of marquesses do not relate to the precedence of any persons then living. This indicates that the precedence of the sons of dukes, earls, and barons was already settled and that the sons of marquesses were intercalated in their appropriate places.

There is nothing in the 'Order' to indicate that it was drawn up by anyone in authority. It was probably no more than someone's statement of what he understood to be the current practice. This 'Order' or some predecessor can therefore fairly be regarded as the lineal ancestor of the modern tables of precedence. Although many items have been added over the centuries, some of which have since become obsolete, its main framework is the basis of the modern tables.

An official 'Order of all States of Worship and Gentry of England' was issued by 'the Lord Protector's Grace' (John, Duke of Bedford) and the Earl Marshal (John, Duke of Norfolk) for the coronation of Henry VI on 6 November 1429.[30] Although described as Lord Protector, Bedford was also Constable of England, and it was probably in this latter capacity that he acted on this occasion, since the Constable and the Earl Marshal were frequently associated, for example, in the Court of Chivalry. This Order contains twenty-five grades of laymen. It is identical with the 1399 'Order' as far as the younger sons of 'Bachelor Knights', save that 'Knights of

---

[29] *C.P.* xi, pt.i.45.
[30] There are seventeenth-century copies in I.T. MS. Petyt 528/44, fo.29; Bodl.MS. Ashm. 857, fos.142-4; and Coll.Arm.MS. Vincent 151, p. 127, the last printed in [Young], *Ancient Tables of Precedency*, pp. 5,6.

the Order of the Counsaile' are omitted and 'the King's Squyres by Prerogatyve' substituted for 'Squires of the King's Chamber'. Finally, the last two items ('Squires and Gentlemen' and 'Gentlemen made by the Kings Grace') are replaced by:
> Gentlemen of the Kings creation
> Squires of Lords and Knights
> Citizens and Burgesses, Artificers and Yeomen, &c

This 'Order' is accompanied by a processional for the coronation. The processional, unlike the 'Order', which is arranged in descending order of precedence, is arranged in ascending order. It includes places for many office-holders, such as the judges, officers of the Royal Household, and the Great Officers of State. The Great Officers of State are divided into two categories. First come the Lord Chancellor, the Lord Treasurer, the Lord President of the Council, and the Lord Privy Seal, who are stated to be above all except the King's children, uncles, and nephews. This must have been copied from some earlier document, since at the time of his coronation Henry VI had no child, uncle, or nephew. The second category comprises the Lord Great Chamberlain, the Constable, the Marshal, the Admiral, the Steward, the Chamberlain and the Secretary, who are stated to have 'places by their offices of all that be of the degree they be of.'[31]

The Constable of England alone issued 'Orders according to Auncient Statutes.... for the placeing of all Estates, as well in Proceeding before the Prince as in other Assemblyes of Parliament, &c' in 6 Edward IV (1466-7).[32] These 'Orders' were not drafted by adopting those of 1399 and 1429, both of which were arranged in descending order, but were based on a processional in ascending order, retaining some of the directions, such as that the judges were to go two and two, and that they were to be preceded by pursuivants of arms. The 'Orders' of 1466-7 are much more detailed than those of 1399 and 1429. Like the 1429 coronation processional, they include places for office-holders, officers of the Royal Household, and the Great Officers of State.[33]

---

[31] I.T. MS. Petyt 538/44, fo.31.
[32] There are seventeenth-century copies in I.T. MS. Petyt 538/44, fos.104-6; Bodl.MS. Ashm. 857, fo.139; and Coll.Arm. MS. Vincent 151, p. 117, the last printed in Young, op.cit., pp. 7,8.
[33] I.T. MS. Petyt 538/44, fos.104-6 conflates the 1466-7 processional with another, which the names of the peers mentioned in it show to have been drawn up in 1572.

Another 'Order of all Estates' is attributed to Anthony, Earl Rivers, though his authority to issue it is not apparent.[34] It is dated 10 June 1479 '20 E.4', but 10 June in the twentieth year of Edward IV was in the calendar year 1480. It is unlikely to have been drawn up as late as 1479 or 1480, since it describes viscounts as 'of late founded', which was hardly appropriate in view of the fact that the first viscount had been created as long ago as 1440. Suspicions as to its authenticity are increased by the ranking of viscounts below the eldest sons of earls and the omission of the younger sons of earls and the younger sons of viscounts.

Of greater authority is the 'Series ordinum omnium Procerum, Magnatum, et Nobilium', laid down by Jasper, Duke of Bedford and 'other Noblemen' with the approval of Henry VII.[35] Bedford was one of the Commissioners for executing the office of High Steward at the coronation of the Queen Consort on 10 November 1487, so this 'Order' was probably drawn up for that occasion. It certainly cannot be later than 21 December 1495, when Bedford died. Coke described it as 'a record of great authority'.[36] However, its authority was not legislative. It did not contain provisions to be observed in the future. It was a statement of the law as the compiler believed it then to be.

The most authoritative pre-Reformation statement of the law of precedence among laymen is the order taken for 'the placynge of Lordes and Ladyes', entitled 'Precedence of Great Estates in their owne degres'.[37] This was drawn up on the occasion of a dinner given to the French and Venetian ambassadors in the Great Chamber at Richmond by the Lord Chamberlain (Charles, Earl of Worcester) in 1520. It was so highly regarded that it was adopted by the Commissioners for executing the office of Earl Marshal in 1595, when they were ordered by Elizabeth I to inquire into place and precedence.[38] It can fairly be described as the basis of the modern system of precedence, which has been produced by making legislative additions to it.

The 'Order' of 1520 was of particular importance in that it finally settled the precedence of viscounts, which had been a matter

---

[34] Printed in [Young], *Ancient Tables of Precedency*, pp. 9-10.
[35] Coll.Arm.MS. Vincent 151. Printed in part in Young, *Privy Councillors and their Precedence*, pp. 46-7. Young omitted the second part headed 'Foeminae'.
[36] 4 Co.Inst. 363.   [37] Printed in Appendix I., pp. 98-100 *post*.
[38] See pp. 24-5 *post*.

General Precedence among Men before 1534

of some confusion for nearly a century. The first viscount was John, Lord Beaumont, who was created Viscount Beaumont on 12 February 1440. By his patent of creation he was granted precedence before all barons of the realm. This placed him below the eldest sons of earls and the younger sons of marquesses.[39] However, by letters patent dated 12 March 1445 Beaumont was advanced before the eldest sons of earls, and placed immediately after earls.[40] Despite this expression of the royal will, in the 'Order' of 1479 viscounts were placed below the eldest sons of earls. On the occasion of the funeral of Edward IV in 1483 a question arose whether Viscount Berkeley should have place of Lord Maltravers, the eldest son of the Earl of Arundel. The question was decided in favour of Maltravers in accordance with the 'Order' of 1479, and the viscounts were accorded the same ranking in the 'Series Ordinum' drawn up by Jasper, Duke of Bedford early in the reign of Henry VII. They were not accorded the precedence given to Viscount Beaumont by the 1445 patent until the 'Order' of 1520 was drawn up, and this they have retained ever since.

The various 'Orders' defined the precedence of the degrees in the peerage. The normal ranking of the peers of each degree among themselves was in accordance with the dates of the creation of their peerages. However, just as the precedence of the degrees in the peerage was derived from the exercise of the royal prerogative, the ranking within the degrees was also subject to the royal prerogative. Thus, when the newly-created John, Earl of Somerset was introduced into Parliament in 1397 he was placed by the King between the Earls of Nottingham and Warwick,[41] and when Henry VIII created Anne Boleyn Marchioness of Pembroke in 1532 it was with precedence over all other marchionesses.[42]

The precedence of knights is first recorded in the 'Order' of 1399.[43] Knights there follow barons' eldest sons. First come knights of 'the Order of the Counsaile', i.e. knights who are members of the King's Council. After them come knights bannerets and knights bachelors, with barons' younger sons between them.

---

[39] 'Order' of 1429: see p. 15 *ante*.   [40] *C.P.R. 1441-6*, p. 348
[41] *Rot.Parl.* iii.343.
[42] *C.P.* x.404. For other instances of creations with special precedence, see ibid. i.App.C., and W. Prynne, *Brief Animadversions on ... the Fourth Part of the Institutes ... by Sir Edward Cooke*, (London, 1669), pp. 324-7.
[43] For this and other 'Orders', see pp. 14-17 *ante*.

There was no essential difference between any of these classes of knights. Knights bannerets differed from the others only in that they were created as a reward for bravery on the field of battle. This practice had been long disused when it was revived for the nonce by George II, who made sixteen knights bannerets on the field of Dettingen in 1743.[44] Knights bannerets created under the royal banner displayed in open war, the King or the Prince of Wales being personally present, had precedence before other knights bannerets.[45] In the 'Order' of 1399 the eldest sons of knights bannerets and knights bachelors followed knights bachelors, and the younger sons were separated from their elder brothers by esquires of the King's Chamber.

This ranking of knights is repeated in the 'Order' of 1429, but in that of 1466-7 it is augmented by the inclusion of knights of the Garter and knights of the Bath. Here the knights of the Garter follow barons' younger sons and are followed by knights of the Privy Council and those of the Bath. Then come the judges and law officers, who are followed by knights bannerets and knights bachelors.

There seems to have been some uncertainty about the precedence of knights in the latter half of the fifteenth century, for in the 'Order' of 1479 knights of the Garter and of the King's Council precede the eldest sons of barons, while in the 'Series Ordinum' of Jasper, Duke of Bedford, they are further advanced to precede the younger sons of earls. The matter was not finally settled until 1612.[46]

The early medieval 'Orders' were concerned with personal rather than official precedence. Although not designed to be a statement of the law of precedence, the list of office-holders in the statute 12 Ric.II, c.2 throws some light on their precedence among themselves. They are set out in the following order: Chancellor, Treasurer, Keeper of the Privy Seal, Steward of the King's House, King's Chamberlain, Clerk of the Rolls, Justices of the King's Bench and Common Pleas, and Barons of the Exchequer. Although justices are frequently mentioned in the salutations of medieval royal charters, these tell us no more than that they ranked below

---

[44] *C.P.* iii.572-3.

[45] Royal Decree of 28 May 1612, printed in F. W. Pixley, *History of the Baronetage* (London, 1900), pp. 20-1.

[46] See p. 40 *post*.

barons. We do not get any further information as to the ranking of office-holders in relation to other persons until the 'Orders' made by John Tiptoft, Earl of Worcester in 1467.[47] Here the judges are placed above knights bachelors and below knights of the Bath.

The authority for the precedence of laymen being the royal prerogative, which is a matter of English law, and ecclesiastical precedence being a matter of canon law, it was inevitable that there should be separate orders of precedence for churchmen and laymen. It was therefore impossible to determine the relative precedence of, say, a bishop and a viscount or an abbot and a baron. Hence the separation of ecclesiastics and laymen in the salutations of royal charters and letters patent, which was often circumvented by the substitution of the formula: 'to all to whom these presents shall come'. This difficulty may also have been the reason why the seating arrangements in Parliament segregated the spiritual peers from their temporal counterparts.[48] It may also be that this method of seating in Parliament was derived from the arrangements in the royal palaces when the Norman kings formally wore their crowns at Christmas, Easter, and Pentecost. Nevertheless, there must have been occasions when the separation of churchmen and laymen would have been inconvenient.

Such an occasion was described in 'The fashion as the young king's grace and other lords went to the Parliament'.[49] The 'young king' is not named and the date of the Parliament is not given, but internal evidence shows that it relates to the opening of Parliament by Henry VI on 12 November 1439.[50] Those taking part in the procession are arranged in the usual manner, i.e. in ascending order of precedence. Reversing the relevant entries, we get:

    Archbishops
    Lord Chancellor and Lord Treasurer
    Dukes of the Blood Royal[51]

---

[47] See p. 16 *ante*.   [48] Powell and Wallis, *House of Lords in the Middle Ages*, p. 556.
[49] I.T. MS. Petyt 538/44, fo.36.
[50] Those present included 'The Lord Cardinal' and 'Duke Humfrey', i.e. Henry Beaufort, Bishop of Winchester, who was a cardinal from 1426 to 1447, and Humphrey, Duke of Gloucester (d.1446). The Archbishop of York (John Kemp) is not described as a cardinal, so the date must be before he became a cardinal in December 1439. The only session of Parliament which fits these dates is that opened in 1439.
[51] John, Duke of Norfolk, the only duke at this time not of the blood royal, was engaged in an embassy to France and so not in the procession.

General Precedence among Men before 1534

      Marquesses
      Earls[52]
      Bishops
      Barons of the Parliament
      Priors and Abbots
      Younger sons of Dukes.

There is no reason to believe that this was the only occasion on which an official procession was so marshalled, but such occasions were probably rare. This is indicated by a table of precedence used by John Russell, the usher and marshal of Humphrey, Duke of Gloucester, who must have taken a professional interest in the procession to the 1439 Parliament. The name of the compiler of this table is not known, but he was obviously in some doubt as to the relative precedence of ecclesiastical and lay dignitaries, for the relevant entries in his table were:

      Archbishop
      Duke
      Bishop
      Marquess
      Earl
      Viscount
      Legate
      Baron
      'Suffrigan'
      Mitred Abbot
      3 Chief Justices and Mayor of London
      Abbot without mitre
      Knight bachelor
      Prior, Dean, Archdeacon, Knight (*sic*)[53]

More reliable than the anonymous table of precedence used by Russell is the practice observed at the court of Edward IV. In the table prefixed to the Black Book of the Household of Edward IV the persons at court are mentioned in the following order: dukes, marquesses, earls, the King's confessor being a bishop, the Chancellor of England, the Chamberlain of England, the two Chief Justices, viscounts, and barons.[54] The corresponding ranking in

---

[52] There was no viscount at this time.
[53] This table is further discussed at pp. 26-7 *post*.
[54] B.L., Harl. MS. 642, fo.5a, b, printed in A. R. Myers (ed.), *The Household of Edward IV* (Manchester, 1959), p. 77.

the Royal Household Ordinance of 1478 is duke, earl, bishop, baron.[55] Since there was no viscount in 1478, this leaves it doubtful whether a bishop followed or preceded a viscount. This doubt was not finally resolved until the latter half of the next century.[56]

[55] Queen's College, Oxford, MS. 134, fo.17a, printed in Myers, op.cit., p. 224.
[56] See p. 49 *post*.

CHAPTER II

GENERAL PRECEDENCE AMONG MEN SINCE THE ACT OF SUPREMACY (1534)

The medieval concept of the realm of England as 'a body politike, compacte of all sortes and degrees of people, divided in termes and by names of spiritualitie and temporalitie', the one governed by the 'lawe devine' and the other by 'the lawes temporal',[1] lasted until the Reformation and preserved, with much else, the dichotomy of the separate precedence of laymen and ecclesiastics. The abolition in 1532 of appeals to the see of Rome, with the requirement that all causes determinable by any spiritual jurisdiction should be adjudged and determined within the King's jurisdiction and authority and not elsewhere, was the first step towards making it possible for the two scales of precedence to be homogenized.[2] The next step was the parliamentary declaration in 1534 of Henry VIII to be the 'only Supreme Head in earth of the Church of England, called *Anglicana Ecclesia*, with all honours, dignities, and preheminences thereto belonging and appertaining'.[3]

This repudiation of papal authority did not automatically abolish the rules of the canon law. Instead, they became part of 'the King's ecclesiastical law of England'.[4] As such they became, like any other part of the law of England, subject to amendment by Act of Parliament. The way was thus paved for the definition of the precedence of ecclesiastical as well as that of lay magnates by the House of Lords Precedence Act 1539.

Both the short title conferred on the Act of 1539 by the Statute Law Revision Act 1948 and the long title, 'An Act for the placing

---

[1] Ecclesiastical Appeals Act 1532, preamble.

[2] Ibid., ss.1, 3.

[3] 26 Hen.VIII, c.1. This Act has no statutory short title, but it has been usually known as the Act of Supremacy. It is now necessary to append the date of the Act since the statute 1 Eliz.I, c.1. was given the short title of 'Act of Supremacy' by the Statute Law Revision Act 1948, sch.2.

[4] *Caudrey's Case* (1591), 5 Co.Rep.1,9. This passage appears not to be part of the decision, but of Coke's comment upon it. Cf. R. C. Mortimer, *Western Canon Law* (London, 1953), p. 59.

of the lords in the Parliament', are somewhat misleading, for it is provided by s.10 that the order of precedence assigned to the Great Officers of State and the King's Chief Secretary in Parliament shall also be observed in 'the Star Chamber and in all other assemblies and conferences of Council', and the Act has been regarded as governing the precedence of these persons generally.

It may be that Henry's aim in promoting the Act was not so much to settle the precedence of the ecclesiastical and lay magnates as to give precedence to his faithful minister Thomas, Lord Cromwell, whom he had appointed to be his Vicegerent in Spirituals on 18 July 1536. This office was an innovation and carried no precedence. By s.2 of the Act the Vicegerent was placed above the Archbishop of Canterbury.[5] After Cromwell's fall in the summer of 1540 the office of Vicegerent was left unfilled, but the Act has remained in full force and effect in so far as it relates to the other 'great men' mentioned in it.

Precedence is defined in the Act of 1539 by reference to the seats to be occupied in the Parliament Chamber, with the Vicegerent and the two archbishops and the bishops on the right as seen from the throne, with the other laymen on the left. Although the previous system of two orders of precedence was continued by the Act, the importance of the Act lay in its definition of ecclesiastical precedence by non-papal authority. The subsequent placing of the archbishops and bishops with laymen in one order of precedence, although an important change from the practical point of view, was therefore not an innovation of a fundamental nature. The fundamental change had been made by the Act of 1539. So far as laymen were concerned, the Act of 1539 supplemented the law as it had developed by the reign of Henry VIII.[6]

During the reign of Elizabeth I there was some doubt among 'personages of great Estate birth and callinge' regarding their precedence. In order to settle the matter the Queen instructed the Lord High Treasurer (Lord Burleigh), the Lord High Admiral

---

[5] Cromwell is shown above the Archbishop of Canterbury in the roll of the lords spiritual and temporal present at the Parliament held in Hilary term 1540 (Coll.Arm. MS. Box 40, no.40), partly reproduced in *Heralds' Commemorative Exhibition 1484-1934* (London, 1936), pl.xxx. The roll shows Cromwell's arms as Earl of Essex, and must therefore have been prepared after the end of Hilary term, between 17 April, the date of his creation, and the following 10 June when he was sent to the Tower.

[6] See p. 17 *ante*.

(Lord Howard of Effingham), and the Lord Chamberlain (Lord Hunsdon), as Commissioners for executing the office of Earl Marshal, to call before them the kings of arms, heralds, and pursuivants, commanding them to produce 'all such their ancient Bookes Rolles and Muniments' relating to the matters in question. Among the documents produced to the Commissioners was the order drawn up in 1520 on the occasion of the Lord Chamberlain's dinner to the French and Venetian ambassadors.[7] Subject to making a small amendment regarding the precedence of barons' daughters, the Commissioners rewrote the statement of 1520 in tabular form and added to it some entries after knights bachelors and presented it to the Queen in the form of an ordinance dated 16 January 1595.[8]

The Commissioners' ordinance is the basis of the present law of precedence. The system has since developed by the conferment upon persons and classes of persons of special precedence defined in relation to existing precedence as being above or below some other person or class of person. These changes can best be dealt with by reference to each person or class of persons affected by them.

A. PERSONAL

(i) *The Sovereign and the Royal Family*

So long as the body politic was divided into two parts, named the spirituality and the temporality, as described in the preamble to the Ecclesiastical Appeals Act 1532,[9] the Pope stood at the head of the order of precedence relating to the spirituality and the King at the head of that relating to the temporality. When in 1534 Henry VIII was declared to be the 'only supreme head in earth of the Church of England' he succeeded to 'all honours, dignities, and preheminences thereto belonging and appertaining' by virtue of the statute 26 Hen.VIII, c.1 and so became the only head of the body politic. This statute was repealed in 1554.[10] The repealing statute was itself repealed in 1558.[11] The statute of 1534 was not among those expressly revived by the statute of 1558, so that Elizabeth I was not declared to be head of the Church of England,

---

[7] See p. 17 *ante*. The order is printed in Appendix I, p. 98 *post*.
[8] The ordinance is printed in Young, *Privy Councillors and their Precedence*, pp. 48-50.
[9] See p. 23 *ante*.   [10] 1 & 2 Ph.& M., c.8, s.12.   [11] 1 Eliz.I, c.1, s.1.

but became instead 'Supreme Governor of the Realm in all spiritual and ecclesiastical things or causes as temporal'.[12] So far as the present topic is concerned, this difference is immaterial, for it was stated that the object of the Act was to restore to the Crown of England the ancient 'superiorities and preheminences' belonging to it, and it was provided by s.7 that any other superiority or preheminence was abolished out of the realm. Thus Elizabeth I and her successors have headed the order of precedence in England in part as the successors of the pre-Reformation kings and in part as the successors of the papacy.

During the first year of the reign of Edward VI, when the royal functions were performed by his uncle, Edward, Duke of Somerset, as Protector of the Realm and Governor of the King's Person, Somerset was granted by letters patent place and precedence by a place next on the right of the 'Siege Royal', notwithstanding the provisions of the House of Lords Precedence Act 1539.[13]

No provision was made in the medieval 'Orders' for the possibility that a Queen regnant might have a husband. At common law the husband of a Queen regnant is an ordinary subject.[14] This gave rise to no difficulty in the cases of Mary I and Mary II. When Mary I married Prince Philip of Spain in 1554, the marriage articles provided that he should be King of England jointly with Mary.[15] When Mary II became Queen in 1688, she was already married, and there was no difficulty regarding the precedence of her husband, Prince William of Orange, for they were simultaneously declared by the Lords and Commons to be King and Queen.[16] There was again no difficulty when Mary II's sister Anne succeeded William III in 1702, for when Anne's husband, Prince George of Denmark, was naturalized by Act of Parliament in 1688 it had been provided that he should have precedence as the first nobleman of England.[17]

The marriage of Princess Charlotte, the heiress presumptive of the Prince Regent, to Prince Leopold of Saxe-Coburg was the next occasion on which similar action became necessary. Shortly

---

[12] Ibid., s.9.   [13] *C.P.R. Edw.VI*, i.217.   [14] 3 Co.Inst. 8.

[15] F. Sandford, *Genealogical History of the Kings and Queens of England* (London, 1707), p. 502.

[16] 1 Wm.& Mar.sess.2, c.2, preamble, reciting resolution of 13 February 1688.

[17] 1 Wm.& Mar.sess.1, c.3 (private); N. Luttrell, *Brief Historical Relation of State Affairs* (Oxford, 1857), i.519.

before the wedding a statute (56 Geo.III, c.13) provided that after the marriage was celebrated it should be lawful for the King to give the Prince such precedence and rank before the Archbishop of Canterbury as he should deem fit and proper, any law, statute, or custom to the contrary notwithstanding. The wedding took place on 2 May 1816, and on the following day a royal warrant was issued by which the Prince was placed immediately after the sons of the King's brothers and sisters.[18]

Shortly before Queen Victoria's marriage to Prince Albert of Saxe-Coburg and Gotha in 1840 a bill was introduced for his naturalization and for giving him precedence next after the Queen. Although the Queen's uncles, the Dukes of Sussex and Cambridge, consented to the Prince's precedence, in the House of Lords it met with heavy opposition led by the Duke of Wellington, and the Act for the naturalization of the Prince had to be passed without any provision as to his precedence.[19] It having proved impossible to give the Prince Consort the statutory precedence which the Queen desired, a royal warrant, dated 4 March 1840, was issued declaring that the Prince should upon all occasions and in all meetings, except where otherwise provided by Act of Parliament, have place and precedence after the Queen.[20]

The royal warrant in favour of the Duke of Edinburgh, dated 15 September 1952, is in similar terms to that of the Prince Consort.[21] Effect has been given to the words 'except where otherwise provided by Act of Parliament' by placing the Duke in the Parliament Roll as the junior duke in accordance with the Act of 1539.

The House of Lords Precedence Act 1539 deals with the precedence of the male members of the Royal Family in a somewhat oblique manner. It is provided by s.2 that no one (except only the King's children) shall sit or have place on any side of the cloth of estate in the Parliament Chamber, and by s.4 that the Lord Chancellor, the Lord Treasurer, the Lord President of the Council, and the Lord Privy Seal, being barons of Parliament or above, shall sit above all dukes, except such as shall happen to be the King's son, the King's brother, the King's uncle, the King's

---

[18] I.41, p. 124.
[19] 3 & 4 Vict., c.2; C. C. F. Greville, *The Greville Memoirs (Second Part)* (London, 1885), i. 259, 263, 265-6, 269-70. Greville also wrote a pamphlet entitled *The Precedence Question* (London, 1840), reprinted in his *Memoirs*, i. 395-406.
[20] I.52, p. 214.   [21] I.81, p. 328.

nephew, or the King's brother's or sister's sons.[22] It thus appears that all the King's sons were given the exclusive privilege of sitting on either side of the cloth of estate, but only those who were dukes were given precedence before the four Great Officers of State specified in s.4.

S.4 of the Act of 1539 was supplemented by Princess Sophia's Precedence Act 1711, which provided that the Electress Sophia of Hanover, her son the Elector of Brunswick Lunenburgh (afterwards King George I), her grandson George, Electoral Prince of Hanover (afterwards King George II), and the heirs of her body, being protestants, should have rank and precedence after the issue of Queen Anne and before the Archbishop of Canterbury, the Great Officers of State, and the dukes and other peers.[23] Although never repealed, it would appear that the Act of 1711 could now apply only to an heir presumptive who was not entitled to precedence under the Act of 1539, for any heir apparent would have such precedence as the son of the Sovereign.

The problem of the placing of the Sovereign's grandsons did not arise until the reign of George I. By a royal warrant dated 13 December 1726 Garter King of Arms was directed to place the names of the King's grandsons, Frederick, Duke of Edinburgh, and William, Duke of Cumberland, on the Parliamentary Roll before that of the King's brother, Ernest, Duke of York.[24] From then onwards it has been the settled practice to place the male members of the Royal Family in the following order: the Sovereign's sons, grandsons, brothers, uncles, and nephews.

Even later to arise was the problem of the precedence of the grandsons of former Sovereigns. When Adolphus, Duke of Cambridge, the seventh son of George III, was succeeded by his son, Prince George of Cambridge, in 1850, Queen Victoria enquired of Lord Chancellor Truro as to her power to confer upon the new Duke of Cambridge precedence above all other dukes. Presumably

---

[22] Coke suggested that 'nephew' in this context meant grandson (4 Co.Inst. 361), but the better view seems to be that the words 'or the King's Brother's or Sister's sons' are merely explanatory of the preceding words 'the King's Nephew' (Sir Charles Young, *Order of Precedence* (1851), p. 15).

[23] By misquoting 'heirs' as 'descendants' C. R. Dodd, *Manual of Dignities, Privileges, and Precedence* (London, c.1845), p. 31 achieved the astonishing statement that the Act of 1711 declared that all the descendants of the Electress Sophia preceded the Archbishop of Canterbury.

[24] I.27, p. 111.

she did this because he was her first cousin and so not among those included in s.4 of the Act of 1539. This point had been dealt with by the Commissioners for executing the office of Earl Marshal in 1595, when they followed the Order of 1520 by placing dukes of the blood royal or of kin to the Sovereign above all other dukes.[25] However, in the outcome the Duke of Cambridge was not only placed in Garter's Roll above all the other dukes, but also above the Archbishop of Canterbury and the Lord Chancellor, despite his not falling within the ambit of s.4 of the Act of 1539.[26] This precedent has been followed in all subsequent similar cases.

The medieval 'Orders' also gave special precedence to the sons of dukes of the blood royal. They all placed the eldest sons of these dukes before marquesses, but they were not consistent in the placing of the younger sons. The younger sons are not mentioned in the 'Orders' of 1399 and 1429; in the 'Order' of 1479 they were placed below the eldest sons of marquesses, while in that of Jasper, Duke of Bedford (*c*.1490) they were placed above the eldest sons of marquesses.[27] The places there assigned to grandsons in the male line of former Sovereigns who are not dukes are thus comparatively humble. Yet such grandsons are entitled to the style of Royal Highness and the title of Prince,[28] and the practice has been to accord to them precedence immediately after the members of the Royal Family falling within the ambit of s.4 of the Act of 1539. As the Clerk of the Privy Council put it when writing in 1840 of Prince George of Cambridge before he succeeded to his father's dukedom, when his position in the Royal Family was the same as that of Prince Michael of Kent:

The practice, however, does not wait upon the right, and is regulated by the universal sense and feeling of the respect and deference which is due to the Blood Royal of England. The Archbishop of Canterbury does not take a legal opinion or pore over the 31st of Henry VIII to discover whether he has a right to jostle for that precedence with the cousin, which he knows he is bound to concede to the uncle, of the Queen; but he yields it as a matter of course, and so uniform and unquestionable is the custom, that in all probability neither the Prince nor the

---

[25] See p. 17 *ante* and p. 100 *post*.
[26] Lord Truro had 50 copies of his advice printed for private friends with the title *On the Precedency of Peers of the United Kingdom*.
[27] For these 'Orders' see pp. 14-17 *ante*.
[28] The style and title were confirmed by letters patent dated 30 October 1917, which also provided that they should not extend to the children of such grandsons other than the eldest living son of a Prince of Wales.

Prelate are [sic] conscious that it is in the slightest degree at variance with the right.[29]

It was provided by a royal warrant dated 30 November 1917 that the grandchildren of the sons of the Sovereign in the direct male line (save only the eldest living son of the Prince of Wales) should have and enjoy on all occasions the style and title enjoyed by the children of dukes,[30] but the warrant contains no provision as to precedence.

(ii) *The Peerage*

There has been no basic change in the precedence of peers and their sons since the Reformation.[31] It was provided by s.7 of the House of Lords Precedence Act 1539 that all dukes, marquesses, earls, viscounts, and barons (other than those holding offices with special precedence under the preceding provisions of the Act[32]) should sit and be placed after their 'ancienty' as it had been accustomed. This was but giving statutory effect to the pre-existing practice. In the case of peers whose peerages were created by patent, this could give rise to no difficulty, but from time to time disputes and difficulties have arisen regarding the relative precedence of peerages deemed to have been created by writs of summons.[33] Even when there was no dispute, the application of the rule as to 'ancienty' has sometimes proved to be difficult.[34]

At first it seems to have been thought that the Act of 1539 had not affected the prerogative power of the Crown to regulate the precedence of peers of the same degree *inter se*.[35] Thus, in 1553 Edward Courtenay was created Earl of Devon with a clause in the patent providing that he and his successors were to enjoy in Parliament as well as in all other places, such place and precedence as any of his ancestors, being Earls of Devon, had ever had or enjoyed.[36] This exercise of the royal prerogative was repeated in several subsequent cases, and went unchallenged until Viscount

---

[29] Greville, *The Precedence Question*, as reprinted in *The Greville Memoirs (Second Part)*, i.401.
[30] i.78, p. 25.
[31] For the precedence of the wives and daughters of peers, see pp. 64-6 *post*.
[32] See pp. 46-8 *post*.
[33] As to such disputes and difficulties, see pp. 102-3 *post*.
[34] A number of such cases are set out in 'Precedency anomalously allowed', *C.P.* i, App.D, 472-4.
[35] See p. 18 *ante*.   [36] *C.P.* i.469.

Wallingford was created Earl of Banbury in 1626 with a clause that he was to have precedence as if he had been the first earl created after the King's accession, thus giving him precedence before six more 'ancient' earls. This was disputed by the House of Lords as being contrary to the Act of 1539, but after protest it was agreed that Banbury should have the precedence for his life, but that it should not pass to his heirs.[37]

The matter came to a head on 5 June 1627 when the barony of Mountjoy was created with precedence before all barons created after the previous 20 May. This put Mountjoy before the Lords Fauconberg and Lovelace. The point was referred to the Lords' Committee for Privileges, who reported on 14 April 1628 that Fauconberg and Lovelace should have place and precedence according to the dates of their patents.[38] The King circumvented this decision by creating Mountjoy Earl of Newport on 5 August 1628. He clearly did not acknowledge himself beaten, for he attempted to grant special precedence in other cases, saying in the patent granted to the Duchess Dudley in 1644 that this was done 'out of our Prerogative Royal which we will not have drawn into dispute'.[39] None of Charles I's successors has followed his example in this respect.[40]

Art.XXIII of the Union with Scotland Act 1706 gives peers of Scotland rank and precedency next and immediately after the peers of the like orders and degrees in England at the time of the Union and before all peers of Great Britain of the like orders and degrees created after the Union.[41] Art.IV of the Union with Ireland Act 1800 gives to the holders of peerages of Ireland existing at the time of the Union (1 January 1801) rank and precedency next and immediately after all the holders of peerages of the like orders and degrees in Great Britain existing at that time and provides that all peerages of Ireland subsequently created

---

[37] Ibid. i.400
[38] Ibid. i.470; ix.348. The decision was cited with approval in *R. v. Knollys* (1694), 1 Ld Raym.10,16.
[39] *C.P.*,i.471.
[40] See also N. H. Nicolas, *Observations on the Clauses containing Grants of Precedency in Patents of Peerage* (London, 1831?).
[41] This provision is ignored in *Kelly's Handbook to the Titled, Landed and Official Classes* (1977 edn.), pp. 80,81,85, where the peers of England and Scotland are arranged together in chronological order.

shall have rank and precedency with the peerages of the United Kingdom so created according to the dates of their creations.[42]

The disclaimer of a peerage under the Peerage Act 1963 does not extinguish the peerage and the precedence attaching to it.[43] Disclaimer is not a term of art in English law, so that it is necessary for a statute which authorizes a disclaimer to specify what its effect will be.[44] It is accordingly provided by s.3(1) (a) of the Act of 1963 that one of the effects of the delivery of an instrument of disclaimer is to divest the person disclaiming of all precedence attaching to the peerage. It is made clear that this divesting of the person disclaiming does not affect the precedence of other persons, since it is expressly provided that, if he is married, his wife will be similarly divested, thus leaving unaffected the precedence derived from the peerage by his sons and daughters.

While no new degrees in the peerage have been instituted since the creation of John, Lord Beaumont, to be Viscount Beaumont in 1440, two additional qualifications for the precedence of a baron have been created by statute.

In order to reinforce the peers qualified to hear appeals to the House of Lords, s.6 of the Appellate Jurisdiction Act 1876 authorized the appointment of duly qualified lawyers to be lords of appeal in ordinary. Every lord of appeal in ordinary, unless he is otherwise entitled to sit as a member of the House of Lords, is by virtue and according to the date of his appointment entitled during his life to rank as a baron by such style as the Crown may be pleased to appoint. The patent by which a lord of appeal in ordinary is appointed is drafted entirely differently from a patent creating a peerage. It does no more than confer the office subject to the provisions of the Act of 1876 and specify the style to be enjoyed by the appointee. Such a patent does not operate to give any style or precedence to the wife or children of the appointee: these matters have been dealt with subsequently by royal warrants.[45] The legal position of a lord of appeal in ordinary is similar

---

[42] Until the matter was settled by the Act of 1800 there was some doubt as to the precedence of Irish peers in England: see [John (Perceval), Earl of Egmont], *The Question of the Precedency of the Peers of Ireland in England* (Dublin, 1739).

[43] Sir Anthony Wagner and G. D. Squibb, 'Precedence and Courtesy Titles', in 89 *Law Quarterly Review* (1973), 362.

[44] e.g. Bankruptcy Act 1914, s.54 (1); Landlord and Tenant (War Damage) Act 1939, s.4(1).

[45] As to the styles and precedence of the wives and children of lords of appeal in ordinary see pp. 66-7 post. The form of the patent of appointment is printed in *C.P.* ii. 180, n. (b).

to that of a bishop having a seat in the House of Lords. Both are Lords of Parliament, but neither has a peerage.

There is no power at common law to create a peerage for life.[46] Power to confer a peerage for life by letters patent was given by s.1(1) of the Life Peerages Act 1958. Since a peerage for life had no previous legal meaning, it was necessary that the Act should specify the incidents of such a peerage. S.1(2) (a) of the Act follows the precedent of the Appellate Jurisdiction Act 1876 by providing that a person on whom a life peerage is conferred is entitled to rank as a baron under such style as may be appointed by the letters patent. A life peerage differs fundamentally from a common-law peerage, which is an incorporeal hereditament.[47] Since it only has the incidents specified by the Act of 1958 and there is no mention in the Act of the wives or children of life peers, provision has had to be made for their styles and precedence by royal warrant.[48]

It is not clear when the sons of peers were first allowed precedence but is must have been before 1399, for they appear in the 'Order of all Estates of Nobles and Gentry' of that year. The eldest sons of peers of all ranks are there assigned precedence different from their younger brothers. Not only do the eldest sons precede the younger sons, but there are ranks between them. Thus, the eldest sons of earls precede the younger sons of marquesses and the younger sons of earls are also preceded by barons. The precedence of the sons of peers set out in the 'Order' of 1399, with the addition of the sons of viscounts from 1440, has continued substantially unaltered. However, the Commissioners for executing the office of Earl Marshal by their 'Ordinance or Decree' dated 16 January 1595 reversed the ranking of viscounts and the younger sons of dukes, assigning the lower place to viscounts.[49] When the order of baronets was created in 1611 some doubt arose as to the precedence of the younger sons of viscounts and barons. This was settled in 1612 by James I, who decided that these younger sons should precede baronets.[50] In 1620 James I

---

[46] *Wensleydale Peerage Case* (1856), 5 H.L.C. 958.
[47] *Nevil's Case* (1604), 7 Co.Rep.33a.
[48] As to the styles and precedence of the wives and children of life peers, see pp. 66-7 *post*.
[49] For the 'Ordinance or Decree', see p. 25 *ante*.
[50] See p. 38 *post*.

decided after 'solemn argument' that the younger sons of earls should have precedence before knights of the Garter.[51]

By the time of Charles II it had become the practice to rank the eldest sons of viscounts, the younger sons of earls, and the eldest sons of barons in commissions after Privy Councillors who were not peers and after judges. This was contrary to the decisions of James I, and on 26 March 1676 it was ordered by the House of Lords that the Clerk of the Crown in Chancery and all other clerks out of who offices commissions issued should take care that the eldest sons of viscounts, the younger sons of earls, and the eldest sons of barons should have the same precedence in the engrossing of commissions as that allowed to them on other occasions.[52]

The eldest son of a peer who is summoned to the House of Lords by a writ of acceleration in his father's barony ranks according to the seniority of the barony and not according to his father's senior peerage. Thus Lord St. John of Basing, the eldest son of the Marquess of Winchester, who was summoned in his father's barony in 1572, was ranked as a baron below Lord Clinton, the eldest son of the Earl of Lincoln.[53]

If the eldest son of a peer dies during his father's lifetime, his eldest son is entitled during the life of the grandfather to the same place and precedence as the eldest son of the peer's degree. If the peer's eldest son had been summoned to the House of Lords in his father's barony, the grandson would succeed to the barony.[54]

Since the lords of appeal in ordinary do not hold hereditary peerages, which are incorporeal hereditaments, they have no heirs, so there is no distinction between their eldest and younger sons, who thus do not qualify for the precedence enjoyed by the eldest or the younger sons of barons. The sons of lords of appeal in ordinary were left without any precedence for over twenty years. They were expressly excluded from the royal warrant of 1876 by which the wife or widow of a lord of appeal in ordinary was granted the rank, style, and precedence of a baroness.[55] It was not until 30 March 1898 that the sons of lords of appeal in

---

[51] Young, *Order of Precedence*, p. 33.   [52] 13 *L.J.* 58a.
[53] Bodl. MS. Ashm. 840, p. 127.
[54] Coll. Arm., Chapter Book, 1680, cited in Young, *Order of Precedence*, p. 27.
[55] See p. 66 *post*.

ordinary (as well deceased as living) were granted rank and precedence immediately after the younger sons of all hereditary barons for the time being and immediately before baronets.[56] This warrant was replaced by another of the same date, which must have been ante-dated. The primary purpose of the second warrant was to give to the sons of lords of appeal in ordinary the style and title enjoyed by the sons of hereditary barons, which had been omitted from the first warrant, but it also defined their precedence more precisely as being immediately after the younger sons of hereditary barons 'created or to be created'.[57]

Like lords of appeal in ordinary, life peers and life peeresses do not hold incorporeal hereditaments, so that their sons also do not enjoy any precedence by virtue of their parents' letters patent. A royal warrant granting precedence to the sons of life peers and life peeresses alone would have had to place them above or below the sons of lords of appeal in ordinary, so the difficulty was got over by replacing the second of the warrants of 1898 by a royal warrant dated 21 July 1958, whereby the sons of lords of appeal in ordinary, life peers, and life peeresses were granted rank and precedence among the sons of hereditary barons in accordance with the dates of appointment of their father as a lord of appeal in ordinary or creation of their father as a life peer or of their mother as a life peeress.[58] This warrant altered the position of the sons of a lord of appeal in ordinary by placing them above the sons of hereditary barons created after their father's appointment. Although no hereditary baron has been created since 1964, there will for some years continue to be sons of lords of appeal in ordinary who were appointed before the creation of some hereditary barons.

Although peers' children who have been legitimated by the Legitimacy Act 1926 or the Legitimacy Act 1959 were accorded the style of legitimate younger children of peers by an Earl Marshal's warrant dated 3 June 1970, they were not granted the corresponding precedence.[59]

---

[56] I. 72, p. 207.
[57] I. 72, p. 241. Although the two warrants bore the same date, they cannot have been contemporaneous: the Earl Marshal's warrant relating to the first was dated 14 May 1898, while that relating to the second was dated the following 6 September.
[58] I. 83, p. 139.
[59] I. 81, p. 218. Children of a void marriage who are to be treated as legitimate by

The grandsons of peers do not appear in any of the medieval 'Orders'. In or shortly after 1590 Robert Cooke, Clarenceux King of Arms, advised that they had no precedence.[60] This remained the position until 18 March 1615, when the eldest sons of the younger sons of peers and their heirs male were placed by the Earl Marshal before the eldest sons of knights.[61] The House of Lords presented an address to the King touching the subject on 7 April 1677. The King gave directions to the Earl Marshal that the eldest sons of the younger sons of peers should precede the eldest sons of baronets during their fathers' life-times.[62]

From the time of Charles II the precedence of peers in Parliament was settled for each session by a roll drawn up by Garter King of Arms at the beginning of each session.[63] In 1966 Garter's Roll was discontinued and replaced by a roll prepared by the Clerk of the Parliaments with Garter's co-operation.[64] This roll is printed in the first volume for each session of Parliament in *Parliamentary Debates (Hansard), House of Lords*. If the correctness of this roll is questioned by a peer, his proper course is to present a petition to the House of Lords for consideration and report by the Committee for Privileges in the same manner as a claim to a peerage. It is the practice for the Committee's reports to be expressed with some caution. The usual form ends with some such words as: 'Saving to the said Lord X and to all other Lords and Peers their Rights and Places upon further and better Authority shewn for the same.'[65]

(iii) *The Baronetage*

In 1611 King James I had in mind the institution of a new hereditary dignity between barons and knights. Those upon whom the

---

virtue of the Legitimacy Act 1959, s.2 and born after the commencement of that Act (29 October 1959) may succeed to a dignity or title of honour and can therefore acquire precedence as the legitimate children of a peer (ibid., s.2(3)).

[60] Bodl. MS. Ashm.840, p. 127.
[61] Young, *Order of Precedence,* p, 51.
[62] 13 *L.J.* 102b, 104b.
[63] Until 1801 Ulster King of Arms used to draw up a similar roll of the peers of Ireland. In 1634 a dispute as to precedence between peers of Ireland was determined by the Lord Chancellor of Ireland and other commissioners appointed by a commission under the Great Seal of England (*Clarence Peerage Case* (1805), Minutes of Evidence, p. 5).
[64] Sir Anthony Wagner, *Heralds of England* (London, 1967), p. 150.
[65] e.g. Report on the precedence of Lord Conyers, 6 July 1798 (41 *L.J.* 622b). For form of petition, see p. 118 *post*.

new dignity was conferred were to be called baronets, and they were to rank before knights.⁶⁶ The King did not, however, give effect to his intention by a general decree or warrant. Instead, the matter was dealt with by the inclusion in each patent conferring a baronetcy a clause providing that the grantee and the heirs male of his body should have place and precedence before all knights, whether of the Bath or bachelors.⁶⁷ Before this was done the two Chief Justices, the Chief Baron, the Attorney-General and the Solicitor-General were consulted, and they resolved that the King had power to grant such precedence.⁶⁸ The patent also provided that if any doubt or question as to the precedence of baronets, their wives, sons, daughters-in-law, or daughters which were not determined by the patent should arise, they should be 'determined and adjudged by and according to such other rules customs and laws (as to ... precedence ...) as other degrees of Hereditary Dignity are ordered governed and adjudged'.⁶⁹

The first patent containing these clauses was issued on 22 May 1611. Since the clause granting precedence before knights said nothing about the relative precedence of baronets and barons, it is not surprising that before the year was out a question had arisen as to the relative precedence of baronets and the younger sons of barons. The question was brought before the Privy Council, which decided against the baronets, who decided to appeal to the King.⁷⁰

The matter was argued by counsel before the King for three days. Before giving his formal decision, the King indicated that he proposed to place the younger sons of viscounts and barons before the baronets, whereupon the baronets petitioned that it should be declared that they should have the very next place to the younger sons of barons without the interposing of any others between

---

⁶⁶ 'A Project for erecting a new Dignitie between Barons and Knights' (P.R.O., SP 14/63/64, printed in Pixley, *History of the Baronetage*, pp. 16-18). Pixley's work is a valuable, though confusingly arranged, corpus of transcripts of documents relating to the baronetage.

⁶⁷ Form of patent, printed in Pixley, op.cit., p. 29. The precedence was also before all knights bannerets, except those created under the royal banner displayed in the royal army in open war, the King being personally present, but this is now obsolete.

⁶⁸ *Anon.* (1611), 12 Co. Rep. 81.

⁶⁹ Form of patent, printed in Pixley, op.cit., pp. 38-9.

⁷⁰ Letter John Chamberlain to Sir Dudley Carleton, 31 Dec. 1611 (P.R.O., SP 14/67/117), printed in Pixley, op.cit., pp. 113-14, where the writer is erroneously described as 'Sir' John.

them.[71] In this the baronets were successful, for by letters patent dated 28 May 1612 the King decreed that the younger sons of viscounts and barons should take place and precedence before baronets, but promised that neither he nor his successors would ever create any other 'degree, order, name, title, rank, dignitie or state' beneath the degree of lords of Parliament which should be higher than or equal to the place of the baronets.[72] Although it was not necessary in order to deal with the precedence of baronets, the King also ordained that the younger sons of viscounts and barons should be placed after knights of the Garter, Privy Councillors, the Master of the Court of Wards and Liveries, the Chancellor and Under-Treasurer of the Exchequer, the Chancellor of the Duchy of Lancaster, the Chief Justice of the King's Bench, the Master of the Rolls, the Chief Justice of the Common Pleas, the Chief Baron of the Exchequer, and the other judges and barons of those courts of the degree of the coif,[73] thus demonstrating that the knights before whom the baronets were given precedence by their patents did not include those of the Order of the Garter. The form of the letters patent creating a baronetcy was accordingly amended so as to incorporate the wording of the decree.[74] Later this was abbreviated to the 'precedences and advantages to the degree of a Baronet in all things duly and of right belonging which other Baronets ... have heretofore honourably and quietly used and enjoyed or as they do at present use and enjoy'.[75]

As between themselves baronets rank in the order of their patents, but the Crown retained a power, which was occasionally exercised, to give a newly-created baronet precedence before one or more of the existing baronets. Thus, on 22 June 1631 Sir Charles Vavasour was created a baronet with place next below Sir Thomas Monson and next above Sir George Gresley, both of whose creations were dated 29 June 1611.[76]

[71] B.L. MS. Cotton Faustina C.VIII, fo.28, printed in Pixley, op.cit., p. 116. See also K. Sharpe, *Sir Robert Cotton* (Oxford, 1979), pp. 124-5.
[72] *The Decree and Establishment of the King's Majestie, upon a Controuersie of Precedence, betweene the yonger sonnes of Viscuntes and Barons, and the Baronets* (London, 1612), also printed in Pixley, op.cit., pp. 119-25.
[73] i.e. those who were serjeants-at-law, thus excluding the Cursitor Baron of the Exchequer.
[74] See a form of the reign of George II, printed in Pixley, op.cit., pp. 32-9.
[75] See a Victorian form, printed ibid., p. 41.
[76] G.E.C[okayne], *Complete Baronetage* (Exeter, 1902), ii. 78. The Vavasour baronetcy became extinct in 1665, but similar creations still affect the ranking of the baronetcies of Goring (1627/1678), Stonhouse (1628/1670), and Acland (1644/1678), and the dormant baronetcy of Stirling-Maxwell (1682/1707).

The institution of the order of baronets in England was followed by the institution of similar orders in Ireland and in Scotland in 1620 and 1625 respectively.[77] The baronets of Ireland and Scotland took precedence in England after those of England until the Acts of Union with Scotland in 1706 and with Ireland in 1800,[78] but since the passing of those Acts the precedence of baronets among themselves has been the subject of some doubt.

There are now five classes of baronets—those of England (1611-1707), Nova Scotia (really Scotland) (1625-1707), Ireland (1620-1800), Great Britain (1707-1800), and the United Kingdom (since 1800). In this respect the baronetage resembles the peerage, but whilst the Union with Scotland Act 1706 and the Union with Ireland Act 1800 make provision for the relative precedence of the classes of the peerage, neither contains any corresponding provision relating to the baronetage. Sir George Young was of the opinion that 'looking at the arrangement in regard to the Peerage', the baronets of Scotland would follow those of England, and the baronets of Ireland follow those of Great Britain, all of whom would be followed by those of the United Kingdom.[79] This opinion is, however, contrary to the rule of statutory interpretation commonly summarised in the maxim *expressio unius est exclusio alterius*.[80] The better view, and that adopted by the successive Registrars of the Baronetage appointed in pursuance of a royal warrant dated 8 February 1910, is that the precedence of baronets among themselves is governed only by the dates of their patents.[81]

From the institution of the baronetage each patent of creation contained special provision for the eldest and other sons of the grantee and of his successors in the dignity. They were to have precedence before the eldest and other sons of all persons before whom baronets had precedence.[82] After the issue of the letters patent of 28 May 1612, by which James I settled the precedence of baronets,[83] the form of the creation patents was amended. The

---

[77] The patents of the baronets of Ireland and Scotland contained clauses relating to precedence worded appropriately for their respective kingdoms: see ibid., pp. 45-6, 83.
[78] Ibid., p. 13.
[79] Young, *Order of Precedence*, p. 45.
[80] Co.Litt. 210a; *Blackburn* v. *Flavelle* (1881), 6 App.Cas. 628, 634.
[81] This is the only matter on which this work differs from the opinion of Garter Young.
[82] Form of patent, printed in Pixley, op.cit., p. 29.
[83] See p. 36 *ante*.

amended form provided that the eldest and other sons of baronets should have precedence before the eldest and other sons of all knights of whatsoever degree or order and also before the eldest and other sons of all persons before whom baronets had precedence.[84] The effect of this amendment was to place the sons of baronets before the sons of knights of the Garter, although baronets were placed by the letters patent of 1612 after knights of the Garter.[85]

The form of baronets' patents was greatly simplified in the early nineteenth century. The specific words concerning the precedence of the sons of baronets were omitted, but their precedence would appear to have been continued by the inclusion of the words: 'All and singular the ... precedences ... to the degree of a Baronet in all things duly and of right belonging.'[86]

(iv) *The Knightage*

During the pre-Reformation period there were but four classes of knights, namely, knights of the Garter, knights of the Bath, knights bannerets, and knights bachelors. The precedence of these four classes between themselves was well-settled, but there was some confusion as to the relative precedence of knights of the Garter and knighted Privy Councillors on the one hand and the younger sons of peers on the other.[87] This confusion continued. In the procession from Somerset House to St Paul's on 24 November 1588 the knights of the Garter and the knighted Privy Councillors had precedence over the younger sons of earls,[88] but in a nearly contemporary account of the opening of Parliament they were ranked below the younger sons of barons.[89] The confusion was not resolved until 28 May 1612, when James I ordained that knights of the Garter and Privy Councillors (other than peers) should have precedence before the younger sons of viscounts.[90]

---

[84] Form of patent, printed in Pixley, op. cit., p. 35.
[85] See p. 38 *ante*.
[86] Form of Victorian patent, printed in Pixley, op.cit., p. 41.
[87] See p. 19 *ante*.
[88] W. Segar, *Honor Military, and Ciuill* (London, 1602), p. 246.
[89] Ibid., p. '251' (*recte* 242). Internal evidence shows that this relates to a parliament between 1575 and 1588, when the Marquess of Winchester attended otherwise than as Lord High Treasurer.
[90] Royal Decree, printed in Pixley, *History of the Baronetage* pp. 121-2. A number of office-holders, including the judges, were placed between Privy Councillors and the younger sons of viscounts.

By an Earl Marshal's Order of 4 February 1626, issued in pursuance of the King's command, knights of the Bath were to have precedence before knights bachelors.[91]

When the Order of the Bath was revived in 1725 it was provided by the statutes that the Knights Companions of the Order who were not entitled to a higher degree of precedency were to enjoy the precedency which had theretofore been allowed any knight of the Bath by ancient laws, usages, and customs of the realm.[92] When the Order was enlarged in 1815 the Knights Companions were renamed Knights Grand Cross and a second rank of Knights Commanders was instituted. The Knights Commanders were given place and precedence 'in all future solemnities and proceedings before knights bachelors'.[93]

By new statutes dated 14 April 1847 the Knights Grand Cross were given precedence next after baronets and the Knights Commanders were given precedence next after the Knights Grand Cross. The current statutes of the Order of the Bath define the precedence of Knights Grand Cross as being next before Knights Grand Commanders of the Star of India and that of the Knights Commanders as being next after the Knights Grand Cross of the British Empire.[94]

This method of defining the precedence of members of orders of chivalry by reference to the precedence of other persons has resulted in a series of interrelated statutes which have to be read together in order to obtain a comprehensive view of such precedence. These are the statutes of the Orders of the Bath, the Star of India, St. Michael and St. George, and the Indian Empire, the Royal Victorian Order, and the Order of the British Empire. The first classes of these Orders come first, followed by the second classes, so that, for example, the first class of the Order of the British Empire ranks before the second class of the Order of the Bath. This ranking of the classes of the Orders, which is reflected in the precedence of the members of those classes, has been slightly modified by a clause in the statutes of the Order of Companions of Honour which provides that that Order shall have precedence

---

[91] Printed in J. Anstis, *Observations .... upon Knighthood of the Bath* (London, 1725), Collection of Authorities, p. 79.
[92] Young, *Order of Precedence*, p. 47.
[93] J. Perkins, *Most Honourable Order of the Bath* (London, 1913), p. 200.
[94] *Statutes of the Most Honourable Order of the Bath 1972*, p. 11.

after the first class of the Order of the British Empire, but confers no personal precedence on the Companions of Honour.[95]

(v) *Esquires*

Despite what James Boswell, writing in 1791, called 'the indiscriminate assumption of Esquire',[96] which has since given rise to the modern practice of affixing 'Esquire' to the name of any untitled man, whether a man is an esquire is a matter of law. This is made clear by the statute of 1392, which provided that no man of lower estate than esquire should wear the livery of any lord if not continually dwelling within the lord's house.[97]

Although whether a man is an esquire is a matter of law, it is a matter on which there is little direct authority. Coke laid it down that the sons of peers and lords of Parliament are in law esquires,[98] but that was far from exhaustive. Who else could be classed as an esquire was a matter of importance at the heralds' visitations in the sixteenth and seventeenth centuries, since they were empowered by commission under the Great Seal to make infamous by public proclamation any person who had 'unlawfully or without just authority, vocation, or due calling' usurped or taken upon him the title of esquire.[99]

Sir William Dugdale, Garter King of Arms, said that in executing such a commission the heralds were to accept as esquires the heirs male of noblemen's younger sons and knights and their descendants, those who could show by long prescription that their lineal ancestors were so styled, sheriffs of counties and justices of the peace so styled in their commissions (during office only), and divers of the King's servants by reason of their offices, such as the heralds and serjeants-at-arms.[100] Blackstone's definition of esquires, culled from another king-of-arms, William Camden,

---

[95] *Statutes of the Order of Companions of Honour 1919*, p. 3. There is no corresponding provision in the statutes of the Order of Merit.

[96] G. B. Hill (ed.), *Boswell's Life of Johnson* (Oxford, 1934), i.34.

[97] 16 Ric.II, c.4. There is a similar provision in 20 Ric.II, c.2.

[98] 2 Co.Inst. 667. Thus, for example, the correct description of a baron's son in a formal document is 'A.B., esquire, commonly called the Honourable A.B.'.

[99] e.g. Commission to Sir Edward Bysshe, Clarenceux King of Arms, dated 7 July 1663, printed in *Hastings Peerage Case* (1840), Minutes of Evidence, p. 187.

[100] Letter Dugdale to Ulster King of Arms, printed in A. R. Wagner, *Heralds and Heraldry* (2nd edn., Oxford 1956), p. 5. The letter is undated, but a passage beginning 'When I visited my province beyond Trent as Norroy' indicates that it was written after Dugdale became Garter in 1677.

Clarenceux, was similar to that of Dugdale, though differing in some details.[101]

The difficulty in defining esquires was recognized by Parliament in 1660. By 'An Act for the speedy provision of money for disbanding and paying off the forces of this Kingdome both by Land and Sea'[102] commissioners were appointed to raise a tax graduated by rank. Among those to be taxed was 'Every person of the degree of an Esquire or so reputed'. This must have been regarded as unsatisfactory, for later in the year an Act for supplying and explaining certain defects in the earlier Act provided that barristers and other persons who had signed deeds in which they were described as esquire or had acted under statutes as esquires should be taxed as reputed esquires.[103]

In *Messor* v. *Molyneux*[104] the Court of King's Bench refused to permit the reading of an affidavit in which a barrister was styled 'gentleman' on the ground that a barrister is an esquire by his office or profession.[105] In *Talbot* v. *Eagle*[106] it was held that the captain commandant of a corps of volunteer infantry was not an esquire and therefore not qualified to kill game by s.4 of the statute 6 Anne, c.16. Although the defendant was styled 'esquire' in his commission signed by the lord lieutenant of Suffolk, this availed him nothing, since it was held that the lord lieutenant had no power to confer honours.[107]

It is not surprising that the precedence of such a miscellaneous and ill-defined class of men as esquires should not be susceptible of precise definition. As Thomas Milles of Gray's Inn, writing in 1610, said: 'Howbeit we see no certainty to be here set down concerning the places of Esquires or their Wiues ... forasmuch as many such things oftentimes chance, as cannot in any certaine rules be comprehended: like as it vseth to happen in *Named Nobility* (viz:) in Princes, Dukes, Marquesses, Earls, Viscounts, and

[101] 1 Bl.Comm. i.406; W. Camden, *Britannia* (London, 1772), i.130.
[102] 12 Car.II, c.9.
[103] 12 Car.II, c.29.
[104] (1741), Unrep., cited in *R.* v. *Brough* (1748), 1 Wils. 244.
[105] An affidavit in which the deponent is described simply as 'esquire' or 'gentleman' will not be received in proceedings in the High Court (*Practice of the Supreme Court 1979*, i.247).
[106] (1809), 1 Taunt. 510.
[107] The defendant also relied on the statute 44 Geo.III, c.54, s.36, by which all officers of corps of volunteers ranked with the officers of the regular forces, but it was held that this meant only the same military rank.

Barons.'[108] Milles's view was re-iterated by Sir Charles Young, Garter King of Arms, in 1851, when he wrote that esquires 'have no peculiar precedency in general society assigned, either by Statute, fixed rule, or ancient usage'.[109]

Notwithstanding this uncertainty regarding the precedence of esquires in general, some esquires have long had a well-defined precedence. These fall into two categories. The sons of peers and peeresses in their own right have various degrees of precedence among themselves, but all of them are above knights bachelors. The other category, consisting of the sons of baronets and knights, all come after knights bachelors. The precedence of this second category of esquires regulates the precedence of a number of other classes of persons below the rank of knight bachelor. This has come about because when the Order of the Bath was enlarged in 1815 an innovation was made by the institution of a third class of members of the Order, to be known as Companions, who were given place and precedence before all esquires of the United Kingdom.[110] Read literally, this would place the Companions above the sons of peers, but since the latter have precedence before knights and the Companions formed the third class in the Order, below the Knights Commanders, it is clear that by 'all esquires' was meant all esquires below the rank of knight.

The constitution of the enlarged Order of the Bath has been the pattern for the orders of chivalry subsequently founded. Each of these orders has included a class or classes of unknighted members to whom has been assigned precedence next after the Companions of the Order of the Bath and next after each other. The Royal Victorian Order was constituted with five classes of members, two knighted and three unknighted, the members of the fourth and fifth classes having precedence next after the Companions of the Distinguished Service Order and the eldest sons of knights bachelors respectively.[111] The precedent of the Royal Victorian Order was followed when the Order of the British Empire was founded in 1917. The members of its fifth class, the Members, have precedence next after the members of the fifth class of the Royal Victorian Order.[112]

---

[108] T. Milles, *Catalogue of Honor* (London, 1610), p. 87.
[109] Young, *Order of Precedence*, p. 59.
[110] Perkins, *Most Honourable Order of the Bath*, p. 203.
[111] *Statutes of the Royal Victorian Order 1936*, art.VII.
[112] *Statutes of the Most Excellent Order of the British Empire 1970*, art.XXII.

Below the younger sons of knights bachelors come the general body of esquires. It is these to whom the observations of Thomas Milles and Sir Charles Young are really applicable. There have, however, been attempts to arrange men below the rank of a knight bachelor's younger son in some sort of order. In 1637 the English heralds advised Thomas Preston, Ulster King of Arms, that a dean out of his diocese had precedence before a doctor of divinity out of his university.[113] Counsel for the defendant in *Ashton* v. *Jennings*[114] admitted that a doctor of divinity took place of an esquire, but said that that was in respect of his degree which was personal to himself and gave his wife no precedence above the wife of an esquire. Wylde J. interjected that doctors of divinity contended with serjeants-at-law for precedence.[115]

Blackstone placed colonels, serjeants-at-law, and doctors (meaning, of course, not the general body of medical practitioners, but those having the degree of doctor in any faculty) between the younger sons of knights and esquires, but he cites only 'the heralds' as authority for this.[116] The true position seems to be that no one below the younger sons of knights bachelors has any precedence for which there is lawful authority.[117]

In some tables of precedence gentlemen find a place below esquires. There is legal justification for this, since whether a man is a gentleman is a question of law, albeit one of some difficulty, which gave rise to much litigation in the Court of Chivalry in the seventeenth century.[118] Even after the Court of Chivalry ceased to be in regular session, the distinction between esquires and gentlemen continued to be observed until well into the nineteenth century, the compilers of directories being among the last to do so. It has, however, long since become obsolete, except for the purposes of some heraldic documents, and a modern master of ceremonies would be ill-advised to attempt to classify guests below the younger sons of knights bachelors into esquires and gentlemen.

---

[113] Bodl. MS. Rawl.D.766, fo.58.
[114] (1674) 2 Lev.133.
[115] This exchange is not recorded in Keble's report of the case.
[116] 1 Bl.Comm. 405.
[117] Therefore the arguments in *The Right of Precedence between Phisicians* [sic] *and Civilians* (Dublin, 1720), which has been attributed to Dean Swift, and the strictures in C. G. Young (?), *Serjeants-at-Law* (1864 ?) on a note relating to the precedence of serjeants-at-law in the 1863 edition of *Burke's Peerage* are devoid of any legal basis.
[118] The matter is fully discussed in Squibb, *High Court of Chivalry*, pp. 171-7.

B. OFFICIAL

(i) *The Great Officers of State*

The Great Officers of State are divided into two categories by the House of Lords Precedence Act 1539. By s.4 the Lord Chancellor, the Lord Treasurer,[119] the Lord President of the Council, and the Lord Privy Seal, if peers, are placed above all dukes other than the Sovereign's son, brother, uncle, nephew, or brother's or sister's sons. By s.5 the Lord Great Chamberlain, the Lord High Constable,[120] the Earl Marshal, the Lord High Admiral,[121] the Lord Steward of the Household, and the Lord Chamberlain are placed after the Lord Privy Seal and 'above all other personages of the same estates and degrees as they shall happen to be'. Round described this as 'obviously a merely arbitrary arrangement, which is based on no historical principle, which confuses the great officers and their deputies, and which interpolates the Admiral in their midst'.[122] Nevertheless, in this respect the Act gave statutory authority to a practice which can be found as early as 1429.[123]

Since 1672 the Earl Marshal has always been a duke. In 1714 the Lord Great Chamberlain was Robert, Marquess of Lindsay. His creation as Duke of Ancaster would by the operation of s.5 of the Act of 1539 have given him precedence over all the existing dukes, including Thomas, Duke of Norfolk, the Earl Marshal. In order to prevent this, it was provided by a private Act of Parliament[124] that the Duke of Ancaster and his successors in the dukedom should have place and precedency among the dukes of Great Britain only according to the date of the letters patent creating the dukedom, except only when in the actual execution of the office of Lord Great Chamberlain, attending the person of the King or Queen for the time being, or introducing a peer into the House of Lords.

The Act of 1714 was still in force when Blackstone published his order of precedence in 1773, so he appended to the entry relating to the Lord Great Chamberlain the note 'But see Private Stat. 1 Geo.I. c.3.'[125] The provision of the Act became spent

---

[119] This office has been in commission since 1714.
[120] This office is now filled only on the occasion of a coronation.
[121] This office has been in commission since 1828.
[122] J. H. Round, *The King's Serjeants & Officers of State* (London, 1911), p. 123.
[123] See p. 16 *ante*.   [124] 1 Geo.I, c.3 (private).
[125] 1 Bl.Comm.405.

when the last Duke of Ancaster died in 1779 and the office of Lord Great Chamberlain passed to two co-heiresses. In spite of this, when Blackstone's table was adapted for use in Burke's *Peerage* in 1826 the substance of his note was reproduced as 'when in actual performance of official duty, statute 1st George I'. In this form the note has been continued in successive editions of Burke, whence it has passed into other publications, giving rise to the mistaken belief that the present (1981) Lord Great Chamberlain, who is a marquess, is entitled to precedence before the Earl Marshal.[126]

It was provided by s.8 of the Act of 1539 that if the Lord Chancellor, Lord Treasurer, Lord President of the Council, or Lord Privy Seal should be under the degree of a baron he should sit and be placed at the uppermost parts of the sacks in the Parliament Chamber. This provision relates only to their places in the Parliament Chamber. S.10 of the Act of 1539 provides that 'as well in all parliaments as in the Star Chamber, and in all other Assemblies and Conferences of Counsel' the Lord Chancellor, Lord Treasurer, Lord President, Lord Privy Seal, Lord Great Chamberlain, Lord High Constable, Earl Marshal, Lord High Admiral, Lord Steward, and Lord Chamberlain are to be placed in such order and fashion 'as is above rehearsed'. Since the precedence in Parliament of the Lord Chancellor, Lord Treasurer, Lord President, and Lord Privy Seal is dealt with specifically in ss.4 and 8 of the Act and depends upon whether they are peers, it seems that the words 'as well in all parliaments' in s.10 must be regarded as surplusage and that the precedence of the persons mentioned in that section outside Parliament does not depend upon whether they are peers.[127] This must have been the view

---

[126] The office of Lord Great Chamberlain is anomalous. In 1902 the Committee for Privileges of the House of Lords resolved that the rights of the co-heiresses who had inherited the office in 1779 were vested in the Earl of Ancaster, the Marquess of Cholmondeley, and the Earl Carrington, in whom therefore the right of selection of a deputy vested, subject to the King's approval; that in the event of those lords not all agreeing, the King might appoint whom he would for the performance of the duties of the office until they should agree; and that according to the precedents the person appointed must not be of inferior degree to a knight. The person selected as the deputy in accordance with this decision has always been one of the co-heirs, with a change at the beginning of each reign. Although but a deputy, the selected co-heir, has been styled and accorded precedence as if he were in fact the Lord Great Chamberlain.

[127] G. Bowyer, *Commentary on the Constitutional Law of England* (London, 1841), p. 694, citing the commission for the trial of Mary, Queen of Scots, where Sir Thomas Bromley, Lord Chancellor was named immediately after the Archbishop of Canterbury.

taken in 1919 when the Speaker's precedence was related to that of the Lord President.[128]

(ii) *Diplomatic Representatives*

By a royal warrant issued on 24 December 1948 the High Commissioners of Commonwealth countries and the Ambassadors of foreign states were given precedence after the Lord Privy Seal in order of their seniority based on the dates of arrival in the United Kingdom. It was provided by the same warrant that a minister of the Crown from a Commonwealth country visiting the United Kingdom should be given precedence before the High Commissioner of that country.[129]

(iii) *The Episcopate*

The medieval scale of ecclesiastical precedence in England was short, comprising only archbishops, bishops, abbots, and priors. After the Reformation it was even shorter, and the problems of ecclesiastical precedence were concerned only with the precedence of archbishops and bishops among themselves and their precedence in relation to laymen.

The relative precedence of the Archbishops of Canterbury and York had long been settled.[130] That of the diocesan bishops was settled by s.3 of the House of Lords Precedence Act 1539, which was probably declaratory of the pre-existing position. First comes the Bishop of London, followed by the Bishops of Durham and Winchester, and then the other bishops 'after their ancients', i.e. in order of seniority of consecration.[131]

There were twenty-six lords spiritual until 1847, when the bishopric of Manchester was established. It was then provided by s.2 of the Ecclesiastical Commissioners Act 1847 that the number of lords spiritual sitting and voting in Parliament should not be increased, and similar provisions have been included in subsequent legislation for the creation of new sees. The diocesan

---

[128] See p. 61 *post*.

[129] I.81, fo.329. The precedence of ambassadors in the seventeenth century was dealt with in *Finetti Philoxenis: Or some choice Observations of Sir John Finett, Knight ... touching the Reception and Precedence ... of Forren Ambassadors in England* (London, 1656).

[130] See p. 10 *ante*.

[131] Milles, *Catalogue of Honor*, p. 61, ranks the bishops according to their dates of election, but this is not in accordance with the statute: a bishop's precedence is not affected by translation to another see, unless it be to London, Durham, or Winchester.

bishops who do not have seats in the House of Lords have precedence in order of seniority of consecration next after their seniors who do.

The Act of 1539 continued the medieval practice of dealing with the precedence of churchmen and laymen in the House of Lords separately, and no provision was made for ranking them together in any order outside Parliament. Such ranking was, however, achieved. It has long been settled that archbishops precede dukes and bishops precede barons. Not only was this not brought about by the Act of 1539,[132] but it did not follow immediately after it. An early attempt at assimilation was made in s.3 of the statute 1 Edw.VI, c.7. This provided that a suit should not abate by the plaintiff's obtaining a dignity or title which changed his style. The dignities in question are set out in the following order—duke, archbishop, marquis, earl, viscount, baron, bishop, knight, justice of either Bench, and serjeant-at-law.[133] This arrangement was short-lived, for after the restoration of papal authority by the statute 1 & 2 Ph.& M., c.8 in 1554 the full form of salutation address in documents under the Great Seal reverted to the pre-Reformation form of archbishops, bishops, dukes, marquesses, etc.[134]

The frequent use of the salutation address to 'all to whom these presents shall come' makes it difficult to establish when clerical precedence was re-assimilated with that of laymen after papal authority was finally repudiated by the Act of Supremacy in 1558, but the ranking had become settled by the early years of the reign of Elizabeth I, as is shown by the salutation address to 'archbishops, dukes, marquesses, earls, viscounts, bishops, barons, and knights' in peerage patents.[135]

It was provided by art.IV of the Union with Ireland Act 1800 that the lords spiritual of Ireland should have rank and precedency next and immediately after lords spiritual of the same rank and degree of Great Britain. This placed the Archbishops of Armagh

---

[132] *Pace* 1 Bl.Comm. 405, where the Act of 1539 is cited as authority for the precedence of archbishops and bishops.

[133] Strangely, *Burke's Peerage* (1953), p.cxcv cites this statute as the authority for placing archbishops before dukes and bishops before barons.

[134] e.g. letters patent of Earldom of Devon 3 September 1553, printed in Sir H. Nicolas, *Report of Proceedings on the Claim to the Earldom of Devon*, App, p.xi.

[135] e.g. letters patent of earldom of Essex, 1572, printed in Milles, *Catalogue of Honor*, p. 36.

and Dublin, in that order, next after the Archbishop of York and the other Irish bishops in the order of their dates of consecration after the last consecrated English or Welsh bishop. However, in 1812 the Bishops of Meath and Kildare, who had precedence before the other Irish bishops, claimed that they were entitled to precedence in England immediately after the Bishop of Winchester. This claim was considered by the Committee for Privileges of the House of Lords, who reported against it, in which report the House concurred.[136] All the Irish archbishops and bishops were deprived of their seats in the House of Lords by s.13 of the Irish Church Act 1869, which preserved the precedence of each archbishop and bishop during his life.

The bishops of Welsh sees lost their precedence when the establishment of the Church in Wales was terminated by s.1(1) of the Welsh Church Act 1914, but by s.2(2) of the Act the precedence of a bishop of the Church in Wales was preserved during his life.

The relative precedence of the two archbishops and the Lord Chancellor seems never to have been determined explicitly. While it has been accepted from time immemorial that the Archbishop of Canterbury precedes the Lord Chancellor, there appears to have been some uncertainty regarding the Lord Chancellor and the Archbishop of York. In Tiptoft's 'Order' of 1467[137] the two archbishops were placed before the Lord Chancellor,[138] and at the opening of Parliament in 1584,[139] yet in the procession from Somerset House to St Paul's on 24 November 1588 the Lord Chancellor was placed between the two archbishops.[140] There seems to be no evidence of any formal decision on the point, but the Archbishop of York has yielded precedence to the Lord Chancellor since the reign of Charles II.[141]

This ranking of archbishops before dukes and bishops before barons has continued unaltered, but it is far from clear how it applies to suffragan bishops appointed under the Suffragan Bishops Act 1534.[142] S.2 of this Act provides that suffragan bishops so

---

[136] 48 *L.J.* 801, 926, 937.
[137] See p. 16 *ante.*
[138] I.T. MS. Petyt 538/44, fo.106. This part of the 'Order' is not printed in Young, *Privy Councillors and their Precedence.*
[139] Milles, *Catalogue of Honor,* p. 66.   [140] Segar, *Honor Military and Ciuill,* p. 45.
[141] Young, *Order of Precedence,* p. 17.
[142] 26 Hen.VIII, c.14. For the short title, see Statute Law Revision Act 1948, sch.2.

appointed shall have such 'prehemyence' as 'Suffragans of this Realm hertofore hathe byn used and accustomed'.[143]

The pre-Reformation suffragan bishops to whom the new-style suffragans were thus equated were the titular bishops who assisted the diocesan bishops.[144] Ten of them and the diocesans whom they had served were still alive in 1534, so the interpretation of s.2 of the Act of 1534 would not have presented the difficulty which at first sight it presents to the modern reader.[145] The pre-Reformation titular bishops ranked in canon law below diocesan bishops and above abbots. The real difficulty is one which would not have seemed important to those alive in 1534. Then the official precedence of churchmen and laymen was still regarded as separate. How, if at all, the precedence of suffragans appointed under the Act of 1534 was regulated in relation to that of laymen is not apparent. The assimilation of the two orders of precedence was still in an indeterminate condition in the reign of Edward VI, and it ceased to matter with the reintroduction of the severance under Mary I. Few suffragans were appointed by Elizabeth I, and the last of them, Bishop Sterne of Colchester, died in 1608. The point then ceased to be of practical moment until the practice of appointing suffragans under the Act of 1534 was revived with the appointment of Edward Parry to the titular see of Dover in 1870.

While it is clear that modern suffragan bishops, like their pre-Reformation counterparts, yield precedence to diocesan bishops, their position *vis-à-vis* barons is by no means equally clear. *Kelly* (1977) places them between diocesan bishops and barons, while other current works of reference by not mentioning them imply that they rank with other bishops in seniority of consecration in accordance with s.3 of the Act of 1539.[146] This section can, however, have no application to suffragan bishops, for it is concerned

---

[143] *Statutes of the Realm* (London, 1817), iii. 510. This part of s.2 is not printed in D. Pickering, *Statutes at Large* (Cambridge, 1763), iv. 340.

[144] See p. 11 *ante*.

[145] Archbishop Cranmer is said to have treated suffragan bishops with contempt, not allowing them to dine with diocesan bishops at his own table, but relegating them to the second or almoner's table with his chaplains (J. Lewis, *Some Account of Suffragan Bishops in England* (London, 1785), p. 12; J. Strype, *Memorials of Thomas Cranmer* (Oxford, 1848) i. 514).

[146] C. R. Dodd, *Manual of Dignities* (London, 1842), p. 45 places English suffragan bishops after the Bishop of Winchester, but the word 'suffragan' is here used in its other sense (see p. 11 *ante*), for in 1842 there had been no suffragans under the Act of 1534 for well over two centuries.

with the seating of bishops in Parliament, where suffragan bishops have no place. The only pre-Reformation indication of the place of suffragan bishops in relation to the laity appears to be the entry 'Suffrigan' below 'Baron' and above 'Mitred Abbot' in John Russell's *Boke of Nurture*.[147] There are, however, obstacles to implicit acceptance of this ranking. The first is Russell's unreliability, indicated by his placing of bishops between dukes and marquesses, and the fact that suffragans find no place in the other versions of the table beginning 'Pope hath no pere'.[148] Furthermore, it has to be borne in mind that at the time when Russell drew up his version of 'Pope hath no pere' there had been no official ranking ecclesiastics and laymen in one order of precedence. When this came to be done in the sixteenth century bishops were placed between viscounts and barons without any distinction between diocesans and suffragans. The correct inference to be drawn from this appears to be that all bishops preceded barons and that suffragan bishops continued by virtue of s.2 of the Act of 1534 to have precedence after diocesans. The part of s.2 of the Act of 1534 relating to precedence was repealed by s.15 (2) (a) of the Dioceses Measure 1978, but since that part of s.2 of the Act of 1534 was merely declaratory of what the law had been immediately before the Act was passed, it would seem that its repeal has not affected the precedence of suffragan bishops, which they would have had even if the Act of 1534 had contained no reference to it.

There is no ancient authority regarding the precedence of retired bishops because bishops did not retire until comparatively modern times. There was no provision for the retirement of bishops until the Bishops (Retirement) Measure 1951. Even so, on retirement a bishop remains in episcopal orders and is still a bishop. Since suffragan bishops have precedence immediately after diocesans and the evidence of the Royal pleasure to be derived from the sixteenth-century letters patent shows that bishops in general precede barons, the precedence of retired bishops can only be after suffragans and before barons.

(iv) *The Judiciary*

The Lord Chancellor is not only the head of the judiciary but

---

[147] 34 E.E.T.S., p. 186.
[148] See p. 21. *ante*.

also one of the Great Officers of State, and it is in this latter capacity that his precedence is governed by s.4 of the House of Lords Precedence Act 1539. That Act does not, however, define the relative precedence of the Lord Chancellor and the archbishops of Canterbury and York, though it is now well settled that the Lord Chancellor's place is between the two archbishops.[149]

Sometimes, instead of appointing a Lord Chancellor, the Sovereign has entrusted the Great Seal to a Lord Keeper. Although appointment as Lord Keeper has been regarded as a lesser dignity than one as Lord Chancellor (some have been promoted from the one to the other, the last being Lord Keeper Henley in 1761), the difference is but slight, for in 1562 it was declared by the statute 5 Eliz.I, c.18 that a Lord Keeper should have the like place and preheminence as of right belongs to the office of Lord Chancellor.

During the sixteenth century there was some uncertainty regarding the precedence of the judges. They had been placed in the 'Orders' made by John (Tiptoft), Earl of Worcester in 1467[150] above knights bachelors and below knights of the Bath. They appeared in the same relative position in the procession from Somerset House to St. Paul's on 14 November 1588, but at the opening of one of the parliaments of Elizabeth I the knights bachelors were placed between the judges and the knights of the Bath.[151] This confusion was not resolved until 28 May 1612, when James I ordained that the Chief Justice of the King's Bench, the Master of the Rolls, the Chief Justice of the Common Pleas, the Chief Baron of the Exchequer, all judges and justices of either Bench, and the barons of the Exchequer of the degree of the coif[152] should have precedence after the Chancellor of the Duchy of Lancaster and before the younger sons of viscounts.[153]

The judges named in the decree of 1612 were augmented in 1813 by the appointment of the Vice-Chancellor of England, who was given by s.4 of the statute 53 Geo.III, c.24 rank and

---

[149] See p. 50 *ante*.
[150] See p. 16 *ante*.
[151] Segar, *Honor Military and Ciuill*, pp. '251' (*recte* 242), 246.
[152] i.e. those who were serjeants-at-law: there was a baron of the Exchequer, known as the cursitor baron, who was not a serjeant.
[153] Letters patent exemplifying royal decree, printed in Pixley, *History of the Baronetage*, pp. 121-2.

precedence next after the Master of the Rolls. In 1841 two additional judges assistant to the Lord Chancellor to be called Vice-Chancellors were appointed. These Vice-Chancellors were given by s.25 of the Court of Chancery Act 1841 rank and precedence next to the Lord Chief Baron of the Exchequer during the continuance in office of the then Vice-Chancellor of England, whose precedence next after the Master of the Rolls was before the Lord Chief Baron. After the death, resignation or removal of the Vice-Chancellor of England the Vice-Chancellors appointed under the Acts of 1813 and 1841 were to have rank and precedence next to the Lord Chief Baron and, as between themselves, according to seniority of appointment. This provision came into effect on the death of Sir Lancelot Shadwell, Vice-Chancellor of England, on 10 August 1850. In 1851 and 1852 provision was made for the appointment of Vice-Chancellors with rank and precedence next after the existing Vice-Chancellors.[154]

Meanwhile, in 1831 the superior judiciary had been further augmented by the establishment by the statute 1 & 2 Will.4, c.56 of the Court of Bankruptcy, consisting of a Chief Judge, three other judges, and six commissioners. The letters patent by which they were appointed assigned to them rank and precedence in all the courts of law and equity. The Chief Judge was placed next after puisne judges of the three superior common-law courts, the other judges after the Chief Judge, and the commissioners after the other judges, the other judges and the commissioners ranking among themselves respectively according to the priority of their appointments.[155] By s.2 of the Act of 1831 the judges of the Court of Bankruptcy or any three of them were to form a Court of Review, and by s.65 of the statute 5 & 6 Vict.c.122 the judges of the Court of Review in Bankruptcy were given rank and precedence next after the judges of the superior courts of Westminster Hall. The Court of Review was abolished in 1847 by the statute 10 & 11 Vict., c.102, s.2 of which preserved the rank and precedence to which the then judges of the Court (Hon. Thomas Erskine and Sir George Rose) were then entitled.

The jurisdiction exercised by the Lord Chancellor in the Court

---

[154] 14 & 15 Vict., c.4, s.2; 15 & 16 Vict., c.80, s.53. The last Vice-Chancellor, Bacon, resigned in 1886.
[155] Young, *Order of Precedence*, pp. 38-9.

of Chancery was transferred to a new Court of Appeal in Chancery by s.5 of the Court of Chancery Act 1854. By s.3 of the Act the judges of the new court were styled Lords Justices and were given rank and precedence next after the Lord Chief Baron of the Exchequer.

The existing superior courts were abolished and the Supreme Court of Judicature, consisting of the Court of Appeal and the High Court of Justice, was constituted by the Supreme Court of Judicature Act 1873, which came into force on 1 November 1875. S.5 of the Act provided that persons thereafter appointed to fill the places of the Lord Chief Justice of England, the Master of the Rolls, the Lord Chief Justice of the Common Pleas, and the Lord Chief Baron and their successors should continue to be appointed with the same precedence as theretofore.[156] Every existing judge who was made a judge of the High Court or an ordinary judge of the Court of Appeal retained his existing rank.[157]

The provisions of the Act of 1873 were repealed and replaced by the Supreme Court of Judicature (Consolidation) Act 1925. The Lord Chief Justice and the Master of the Rolls retain their precedence next after the Chancellor of the Duchy of Lancaster under the patent of 1612, which has not been abrogated by the subsequent legislation, and the President of the Family Division has rank and precedence next after the Master of the Rolls.[158]

Lords justices of appeal, if not entitled to precedence as peers or privy councillors, rank after the *ex officio* judges of the Court of Appeal and according to the priority of the dates on which they became judges of that Court.[159] Judges of the High

---

[156] The titles of Lord Chief Justice of the Common Pleas and Lord Chief Baron were abolished by an Order in Council made 16 December 1880

[157] Supreme Court of Judicature Act 1873, s.11.

[158] Supreme Court of Judicature (Consolidation) Act, 1925, s.16(2); Administration of Justice Act 1970, s.1(1).

[159] Supreme Court of Judicature (Consolidation) Act 1925, s.16(3). The *ex officio* judges of the Court of Appeal are the Lord Chancellor, any person who has held the office of Lord Chancellor, any lord of appeal in ordinary who at the date of his appointment would have been qualified to be appointed an ordinary judge of the Court, or who was at that date a judge of the Court, the Lord Chief Justice, the Master of the Rolls, and the President of the Family Division (ibid., s.6(2); Administration of Justice Act 1970, s.1(1)). There is power to appoint a Vice-President of the Court of Appeal, with rank and precedence next after the President of the Family Division (Supreme Court of Judicature (Consolidation) Act 1925, s.16(2A), as amended by Administration of Justice Act 1970, sch.2, para.6), but none has been appointed. As to lords of appeal in ordinary, see p. 32-3 *ante*.

Court who are not also judges of the Court of Appeal rank next after the judges of the Court of Appeal and among themselves according to the priority of the dates of their appointments.[160]

The Lord Chancellor is empowered by s.5(1) of the Administration of Justice Act 1970 to nominate one of the puisne judges for the time being attached to the Chancery Division of the High Court to be Vice-Chancellor, responsible for the organization and management of the business of the Division, but the Act confers no special precedence on the judge so nominated.

When the system of county courts was set up by the County Courts Act 1846 no provision was made for the precedence of the judges. They remained without any precedence as such judges until 4 August 1884, when by royal warrant they were granted precedence next after knights bachelors.[161] By a royal warrant of 17 December 1937 the Official Referees to the Supreme Court appointed under s.125 of the Supreme Court of Judicature (Consolidation) Act 1925 were also granted precedence next after knights bachelors, which had the effect of placing them above County Court judges.[162] The Official Referees and the County Court judges and the holders of certain other judicial offices became Circuit judges by virtue of para.1 of Schedule 2 to the Courts Act 1971. By a royal warrant dated 29 March 1972 Circuit judges were granted precedence next after knights bachelors. As between themselves, Circuit judges rank by virtue of the 1972 warrant in the following order:

> Vice-Chancellor of the County Palatine of Lancaster.
> Circuit Judges who immediately before 1 January 1972 held office as Official Referees according to priority of appointment
> Recorder of London
> Recorders of Liverpool and Manchester according to priority of appointment

---

[160] Supreme Court of Judicature (Consolidation) Act 1925, s.16(4). The ranking of a judge of the High Court is not affected by the Division (Chancery, Family, or Queen's Bench) of the Court to which he is for the time being assigned. The Supreme Court now consists of the Court of Appeal, the High Court, and the Crown Court (Courts Act 1971, s.1), but this reorganization has not affected the precedence of the judges.

[161] I. 68, p. 66. By a royal warrant of 31 July 1968 salaried whole-time chairmen and deputy chairmen of courts of quarter sessions were granted the style and title of 'His Honour Judge - '. but no precedence (I.83,p. 201). These courts were abolished by the Courts Act 1971, s.3.

[162] I. 81, p. 281.

Common Serjeant
Circuit Judges who immediately before 1 January 1972 held the following offices
> Additional Judge of the Central Criminal Court
> Assistant Judge of the Mayor's & City of London Court
> County Court Judge
> Whole-time Chairman or whole-time Deputy Chairman of courts of Quarter Sessions for Greater London, Cheshire, Durham, Kent and Lancashire

according to the priority or order of their respective appointments to such offices
Other Circuit Judges according to the priority or order of their respective appointments.[163]

In some modern tables of precedence there is an entry for Masters in Chancery.[164] Despite the fact that Masters in Chancery were abolished by s.1 of the Court of Chancery Act 1852, it is still necessary to have regard to their precedence. By s.2 of the statute 8 & 9 Vict.,c.100 the Commissioners in Lunacy were renamed Masters in Lunacy and were to take the same rank and precedence as the Masters in Ordinary of the High Court of Chancery.[165] By s.6 of the statute 18 & 19 Vict.c.13 the two Masters in Lunacy appointed under that statute were to have the same rank and precedence as the then Masters in Lunacy. The Masters in Lunacy were continued as theretofore by s.111(1) of the Lunacy Act 1890, but s.1(1) of the Lunacy Act 1922 provided that there should be a single Master in Lunacy instead of two. Finally, the single Master in Lunacy became known as the Master of the Court of Protection when the Office of the Master in Lunacy was renamed the Court of Protection by the Patients' Estates (Renaming of Master's Office) Order 1947.[166]

---

[163] I. 84, p. 79.
[164] e.g. *Dod's Parliamentary Companion* (Herstmonceux, 1980), p. 270.
[165] Masters in Chancery in Ordinary (hereafter referred to simply as 'Masters in Chancery') are to be distinguished from Masters in Chancery Extraordinary, who were empowered to take Chancery affidavits outside London and were usually solicitors or attorneys.
[166] S.R. & O. 1947, No. 1235. The table in *Dod* includes Masters in Lunacy, so ignoring the changes in 1922 and 1947.

The Master of the Court of Protection thus has the precedence enjoyed by the Masters in Chancery in 1845. It is, however, not at all clear what that precedence was. In the reign of Elizabeth I Lord Ellesmere placed the names of Masters in Chancery in commissions before those of serjeants-at-law contrary, so it was said, to 'former precedents'. This ranking was confirmed by the Commissioners for executing the office of Earl Marshal in 1604.[167] However, the controversy was renewed in 1625 on the occasion of the funeral of James I.[168] The question seems to have remained undetermined at that time, but in the processions at the coronations between 1661 and 1821 the serjeants-at-law invariably had precedence before the Masters in Chancery.[169] This, if it settled anything, only settled the relative precedence of serjeants-at-law and Masters in Chancery and throws no light on the rank and precedence of the latter referred to in the statute 18 & 19 Vict.,c.13. The true position seems to have been that both had rank and precedence in the courts of law and that neither had any settled place in the order of general precedence.[170]

(v) *Privy Councillors*

The medieval 'Orders' did not assign any precedence to members of the Privy Council as a body, but in the 'Order' of 1399 [171] 'Knights of the Order of the Counsaile' are placed above knights bannerets. No other members of the Council are mentioned in this 'Order', presumably because all the members at that time were either peers or knights, and it was not necessary to mention peers because they had no precedence by virtue of membership of the Council. By the time of Henry VII the membership of the Council had widened, so that three classes of members below the rank of baron appear in the 'Series Ordinum' of that reign.[172]

---

[167] Young, *Order of Precedence*, p. 61.

[168] Ibid., p. 65.

[169] The relevant part of each processional is printed ibid., pp. 67-9. There was no procession from Westminster Hall to the Abbey at the coronations of William IV and Queen Victoria.

[170] Young, op.cit., pp. 61, 71. This is reflected in such a discrepancy as that between T. Wotton, *English Baronetage* (London, 1727), iii. 438, where Masters in Chancery are placed above knights bachelors, *The Royal Kalendar* (London, 1810), p.ii, where they do not appear at all, and *Dod's Parliamentary Companion 1980*, where they can still be found between Companions of the Distinguished Service Order and Members of the Fourth Class of the Royal Victorian Order.

[171] See p. 14 *ante*. [172] See p. 17 *ante*.

They are knights, who are placed between knights of the Garter and the younger sons of earls; doctors, who are placed between the younger sons of barons and knights bachelors; and esquires, who are placed between knights bachelors and the eldest sons of knights bannerets.

It is not clear how long this separate precedence for knights, doctors, and esquires continued, but all Privy Councillors (other than peers) were dealt with together by a decree of James I dated 28 May 1612. The principal object of this decree was to settle the precedence of the newly-created order of baronets, but that end was achieved by defining the precedence of baronets by reference to that of other persons. In the list of those who were to precede baronets were all members of the Privy Council, who were placed after knights of the Garter.[173] Sir Julius Caesar, remarked that this indirect method of assigning precedence was called by lawyers transitive or enunciative, not dispositive.[174]

Privy Councillors under the degree of the eldest son of a baron rank according to the order of their being sworn members of the Council.[175]

(vi) *Other Office-Holders*

In addition to the Great Officers of State, some other office-holders have precedence by virtue of their offices. Many of them are members of the Government, but members of the Government do not enjoy any precedence as a class. Even the Prime Minister had no precedence as such until by a royal warrant of 10 December 1905 he was granted precedence next after the Archbishop of York.[176] Some members of the Government are entitled to precedence as Privy Councillors, while others have temporary precedence by virtue of the offices which they hold. Thus the Lord Chancellor, the Lord President of the Council, and the Lord Privy Seal rank in accordance with the House of Lords Precedence Act 1539,[177] while the precedence of other members of the Government has been the subject of special provisions.

---

[173] Letters patent exemplifying decree printed in Pixley, *History of the Baronetage*, pp. 121-2; Young, *Privy Councillors and their Precedence*, pp. 17, 25.
[174] Young, op.cit., p. 25.
[175] Ibid., p. 29.
[176] I. 74, p. 286.
[177] See p. 46 *ante*.

The modern Secretary of State is the successor of the King's Chief Secretary of Tudor times.[178] His precedence depends upon whether he is a peer. It is provided by s.5 of the House of Lords Precedence Act 1539 that if the King's Chief Secretary is a baron, he is to be placed before all barons other than those holding the offices mentioned in ss.4 and 5 of the Act.[179] If he is of a higher rank in the peerage than a baron, being Secretary of State does not entitle him to any precedence before other peers of his own degree.[180]

It was provided by s.10 of the House of Lords Precedence Act 1539 that a Secretary of State who was not a baron should have precedence after the Lord Chamberlain.[181] This was, however, effective for only a few months, for when Thomas Wriothesley and Ralph Sadler were jointly appointed the King's Principal Secretaries in March 1540 the royal warrant, besides defining their duties, defined their precedence by providing that the lords, 'both of the temporalty and clergy', and the Treasurer, Comptroller, Master of the Horse, and Vice-Chamberlain should sit above the Secretaries, as well in the King's Household as in the Star Chamber and elsewhere, and that the Secretaries should be next after them.[182]

The office of Master of the Horse has since the reign of Elizabeth I almost invariably been held by a peer, who had his precedence in that capacity until by a royal warrant dated 6 May 1907 he was given precedence next after the Lord Chamberlain.[183] This warrant is a somewhat remarkable document, for the precedence of the Lord Chamberlain is not absolutely fixed, but depends upon his degree in the peerage.[184] Therefore, should the Master of the Horse be of a higher degree in the peerage than the Lord Chamberlain, his precedence would be that of his peerage, but should he be of the same or a lower degree than the Lord Chamber-

---

[178] For an account of the evolution of the office of Secretary of State, see J. R. Tanner, *Tudor Constitutional Documents* (Cambridge, 1951), pp. 202-4.

[179] As to these offices, see p. 46 *ante*. If a Secretary of State is a bishop, he is to be placed before all bishops not having any of the offices mentioned in ss.4 and 5 of the Act of 1539, but no bishop has held this office since 1553.

[180] 4 Co. Inst. 363, citing the case of Robert (Cecil), Earl of Salisbury. Cecil was appointed Principal Secretary of State in 1596 and created a baron in 1603, a viscount in 1604, and Earl of Salisbury in 1605. (*C.P.* xi. 403).

[181] For the interpretation of s.10 of the Act of 1539, see p. 47 *ante*.

[182] *State Papers during the Reign of Henry the Eighth* (1830-52), ii. 633.

[183] I. 75, fo. 109. [184] See p. 46 *ante*.

lain, he would have precedence before all the other peers of the same degree as the Lord Chamberlain.

The precedence of the other three officers of the Household, namely the Treasurer, the Comptroller, and the Vice-Chamberlain has remained as it was defined by the royal warrant of March 1540.

The Chancellor of the Exchequer and the Chancellor of the Duchy of Lancaster were declared by a decree of James I in 1612 to have precedence next after Privy Councillors,[185] but the holders of these offices are usually members of the Privy Council and so have higher precedence in that capacity.

With the exception of the Speaker, members of the House of Commons have no precedence as such, though some of them as members of the Government hold offices to which precedence is attached and some are entitled to precedence as Privy Councillors. In the fifteenth century the Speaker was placed by John Russell in his *Boke of Nurture* next after the Chief Justices.[186] However, the earliest official recognition of the place of the Speaker appears to be in the Act for enabling Lords Commissioners of the Great Seal to execute the office of Lord Chancellor or Lord Keeper. This provided that the Lord Commissioners, if not peers, should have and take place next after peers and the Speaker of the House of Commons,[187] thus indicating that the Speaker's place was immediately after the barons. The precedence of the Speaker was finally settled by an Order in Council of 30 May 1919, under which he has precedence immediately after the Lord President of the Council on all occasions and in all meetings except when otherwise provided by Act of Parliament.[188]

---

[185] Letters patent exemplifying royal decree, printed in Pixley, *History of the Baronetage*, pp. 121-2.

[186] 32 E.E.T.S. p. 188.

[187] Great Seal Act 1688, s.2. From 1690 to 1693 Sir John Trevor was both Speaker and one of the Lords Commissioners and so had precedence before the other Commissioners, neither of whom was a peer.

[188] I. 78, p. 120.

CHAPTER III

GENERAL PRECEDENCE AMONG WOMEN

Although the orders of precedence among churchmen and laymen have been combined into one since the sixteenth century, the order of precedence among women has continued to be treated separately, a rare and possibly the only exception to this practice being the listing of the names of members of the Royal Family in the Book of Common Prayer by Order in Council in accordance with the requirements of s.21 of the Act of Uniformity 1662.[1]

So firmly established was the notion that the precedence of women was entirely different from that of men that when England had a queen regnant for the first time in 1553 it was thought necessary to have a statute declaring that all 'preheminences' belonged to a queen as well as to a king, lest 'malicious and ignorant persons' might be induced and persuaded into 'error and folly'.[2]

Precedence among women is a subject which some have treated from a male chauvinist point of view. Coke said: '... the contention about precedency between persons of that sex is ever firery, furious, and sometimes fatall'.[3] Equally unkind was Guy Miege's sweeping assertion: '*Note*, that all Knights Wives bear the Title of Lady, which makes ambitious Women, being fond of a Title that gives 'em a Precedency over Esquires and Gentlemens Wives, value themselves and their Husbands the more.'[4]

Mrs Ashton of the leading case of *Ashton* v. *Jennings*[5] was fortunate in that she did not live long enough to see William Stephens's satirical tract *Mrs.Abigail, or an Account of a Female Skirmish between the Wife of a Country Squire, and the Wife of a Doctor*

---

[1] Now repealed and replaced by the Church of England (Worship and Doctrine) Measure 1974, s.1(7).
[2] Queen Regent's Prerogative Act 1554.
[3] 4 Co.Inst. 363.
[4] G. Miege, *Present State of Great Britain* (London, 1707), p. 262.
[5] (1674) 2 Lev.133. See p. 1 *ante*.

in Divinity. With Remarks thereupon: containing some free Thoughts on the pretended Quality and Dignity of the Clergy. In a Letter to a Person of Quality, published in 1700. This was clearly based on the case of *Ashton* v. *Jennings.* Stephens, however, embroidered the facts of that case by making his doctor of divinity a domestic chaplain, who had married a lady's maid 'according to the laudable Custom where the Patron had been beforehand with the Chaplain'.[6] Mrs Ashton was in fact no abigail, being a daughter of Robert Warren, rector of Rame, co. Cornwall, a member of the armigerous family of Warren of Hadbury in Ashburton, co. Devon.[7]

The separate treatment of the precedence of men and women goes back at least as far as Domesday Book. In the list of tenants-in-chief at the beginning of the record for each county the names of the women appear after those of the men. This does not indicate that the precedence of women was lower than that of men any more than the listing of churchmen before laymen indicates that all churchmen had precedence before all laymen. The women tenants-in-chief were similarly divided into two groups, the religious preceding the lay women.

There is but little information regarding the precedence of women among themselves to be gleaned from Domesday Book. The only religious women among the tenants-in-chief were the abbesses, and the lay women in any one county were so few that it would be unsafe to infer anything from the order in which they are named, save that countesses came first and possibly that married women and widows preceded spinsters.

While it may be inferred that married women took precedence among themselves according to the ranks of their husbands, the earliest definite statement of the precedence of women is in 'The order for Ladyes and Gentlewomen', which forms the second part of the 'Order' of 1399.[8] There, as in the later 'Orders', the precedence is entirely derivative, depending on the precedence of husbands and fathers.

As in the case of men, the authoritative statement of the precedence of women is to be found in the document entitled 'Pre-

---

[6] *Mrs. Abigail* was reprinted in Stephens's *An Account of the Growth of Deism in England* (London, 1709), pp. 452-67.

[7] *Annual Reports and Transactions of the Plymouth Institution (1887),* 267-8; *Visitation of ... Devon ... 1620* (Harl.Soc. vi, 1872), 299-300, 354.

[8] I.T.MS. Petyt 538/44, fo.29, printed in [Young], *Ancient Tables of Precedency,* p.4: see p. 14 *ante.*

cedence of Great Estates in their own degres', drawn up in 1520.[9] This, like its predecessors, is constructed on the basic principles which continued to be applied, namely that married women are entitled to the same rank among each other as their respective husbands would have between themselves, and that all daughters have the same rank as their eldest brothers would have among men during the lives of their fathers. The only subsequent changes have been those made necessary by the introduction of new ranks, such as baronets, among men. These changes have been made in accordance with the basic principles upon which the 'Order' of 1520 was constructed.

While it is not expressly stated in the medieval 'Orders', it has long been recognized that a wife does not lose her precedence on the death of her husband, nor does she lose it by divorce.[10] If she is a peeress by marriage she retains her title and the privileges of peerage until she marries a man who is not a peer.[11] Even if she marries a peer, she loses the title and precedence derived from her former husband. This made it necessary in 1632 for Charles I to issue a royal warrant declaring that notwithstanding her remarriage to James, Earl of Abercorn, Catherine, Duchess of Lennox, the widow of Esmé, Duke of Lennox, should retain both the title of Duchess of Lennox and the rank and precedence thereunto due.[12]

However, although the widow or divorced wife of a peer loses her rights as a peeress on remarriage, the retention of her previous name and style was held not to be unlawful in *Cowley* v. *Cowley*,[13] where after a divorce the Countess Cowley had married Mr R.E. Myddleton.[14] This decision, however, related only to the

[9] See Appendix I, pp. 98-100 *post*.
[10] *Cowley (Earl)* v. *Cowley (Countess)*, [1901] A.C. 450.
[11] *Countess of Rutland's Case* (1605), 6 Co. Rep. 52b.
[12] I.25, fo.62; *C.P.* i.3; vii. 608. The only other royal warrants of this king appear to have been those granted in 1660 to Rachael, Countess of Middlesex, to retain her precedence as Countess of Bath (her second husband's earldom being junior to that of her first husband) (*C.P.* viii.693), and the licence granted in 1682 to Sarah, Duchess of Somerset to enjoy precedence as a duke's widow notwithstanding any marriage she might thereafter contract (I.25, p. 226v.; *C.P.* xii, pt. i,76). The latter is accordingly described on her monument in Westminster Abbey simply as Duchess of Somerset without any mention of her surviving husband, Henry, Lord Coleraine.
[13] [1901] A.C. 450.
[14] A similar recent instance is that of the widow of the fourth Marquess of Dufferin and Ava, who married secondly Major Desmond Buchanan and thirdly Judge Maude, Q.C.

use of the name and style. Lord Macnaghten was at pains to point out while there was nothing in the Countess Cowley's position as the divorced wife of Earl Cowley to deprive her either of the title or the privileges which she had acquired as his wife, when she married a commoner she lost her right to the title of Countess and the privileges of peerage. Her subsequent use of the title was 'merely a matter of courtesy, and allowed by the usages of society'.[15]

It may be that the rule regarding the loss of precedence on remarriage applies only to peeresses, for Robert Cooke, Clarenceux King of Arms, advised that the widow of Sir Walter Devereux (d.1591), the second brother of the Earl of Essex, retained her precedence after remarriage to Thomas Sidney.[16] There seems to be no direct authority on this point, but it was not uncommon for the widow of a knight or a baronet to retain her previous style and name after marriage to a second husband of lower rank than her first. For example, the widow of Sir William Forth, who died in 1641, was described on her monument at Stowmarket, co. Suffolk as Lady Forth, despite having had esquires as her second and third husbands.[17] It does not, however, follow that Lady Forth and others in the like position continued to enjoy the precedence which they had from their first husbands. The true position may not have been in accordance with Robert Cooke's advice in the case of Lady Devereux. Indeed, it seems more likely that Guy Miege was right when he wrote: 'A Knight's Widow, marrying below her self, is still called *Lady* by the Courtesy of *England,* with the Surname of her first husband'.[18]

A peeress whose husband disclaims his peerage under the Peerage Act 1963 is divested of all titles and precedence attaching to the peerage.[19] There is, however, nothing in the Act to deprive the daughter of a disclaiming peer of her precedence as a peer's daughter.

Although married women in general take their precedence from their husbands, there are two exceptions to this rule. In the first place, a married woman who is entitled to personal precedence as

---

[15] [1901] A.C., at p. 455.
[16] Bodl. MS. Ashm. 840, p. 127.
[17] J. Le Neve, *Monumenta Anglicana ... 1600 to ... 1649* (London, 1719), pp. 194-5.
[18] Miege, *Present State of Great Britain,* p. 289.
[19] Peerage Act 1963, s.3 (1)(a).

a peeress in her own right or as the daughter of a peer enjoys that precedence independently of her husband unless he is a peer. For example, the daughter of an earl married to a knight retains her precedence as an earl's daughter above the wives of other knights, but if married to a baron ranks only as the wife of a baron. But a peeress in her own right married to a peer retains the precedence of her own peerage unless her husband's peerage is senior to her own either by degree or by priority of creation. In the second place, a married woman takes no precedence from any official rank, such as that of one of the Great Officers of State, to which her husband may be entitled.

The exclusion of the wife of an office-holder from the precedence enjoyed by her husband *virtute officii* extends to the wives of bishops. Serjeant Saunders *arguendo* in *Ashton* v. *Jennings*[20] may have appeared somewhat ungallant when he said that a bishop's wife was 'no lady', but he made his meaning clear when he added 'though her husband himself be a Baron of Parliament, and takes place as such'. Even if Queen Elizabeth I had not been so bitterly opposed to episcopal marriage, the position would have been no different, for the rule applies to the wives of all office-holders, whether ecclesiastical or lay.

The wives of lords of appeal in ordinary could have been left in the same plight as the wives of bishops. As already explained,[21] a lord of appeal in ordinary is the holder of an office created by statute and, although by s.6 of the Appellate Jurisdiction Act 1876 he is entitled to rank as a baron, he is not a baron, and therefore his wife is not a baroness. It was therefore provided by a royal warrant dated 22 December 1876 that the wife of a lord of appeal in ordinary whose husband was not otherwise entitled to sit as a member of the House of Lords should be entitled so long as she continued his wife or remained his widow to the rank, style, and precedence of a baroness.[22]

No provision was made by the Life Peerages Act 1958 for the precedence of the wives of life peers, who were thus left in the

---

[20] (1674), 2 Lev. 133.   [21] See p. 32 *ante*.

[22] I.65, p. 331. It was provided that this should not be deemed or construed to authorize or permit any of their children to assume or use the title of 'Honourable' or to be entitled to the style, rank, or precedence of the children of a baron. The children remained subject to this proviso until 1898: see p. 34 *ante* and p. 67 *post*. *C.P.*, ii. 181 incorrectly states that the wives of lords of appeal in ordinary are entitled to rank as baronesses by virtue of the Appellate Jurisdiction Act 1876.

same position as the wives of lords of appeal in ordinary before the royal warrant of 22 December 1876. Since a similar royal warrant in favour of the wives of life peers would have to place them above or below the wives of lords of appeal in ordinary, the 1876 warrant was revoked and replaced by a royal warrant of 21 July 1958. The new warrant granted to the wife or widow of a lord of appeal in ordinary or of a life peer the same style, rank, and precedence as the wives or widows of hereditary barons of the United Kingdom in accordance with the date of appointment of her husband as a lord of appeal in ordinary or his creation as a life peer.[23]

The precedence of daughters derived from the rank of their fathers was extended to the daughters of lords of appeal in ordinary by a royal warrant of 30 March 1898, which gave to the children of lords of appeal in ordinary the style and title enjoyed by the children of hereditary barons and rank and precedence next to and immediately after the younger children of hereditary barons created or to be created.[24] This warrant was revoked by the royal warrant of 21 July 1958, which gave to the children of a lord of appeal in ordinary, a life peer, or a life peeress style and title as the children of hereditary barons with rank and precedence among the children of hereditary barons in accordance with the date of appointment of their father as a lord of appeal in ordinary or the creation of their father as a life peer or of their mother as a life peeress.[25]

The granddaughters of peers do not appear in any of the medieval 'Orders'. In or shortly after 1590 Robert Cooke, Clarenceux King of Arms, advised that they had no precedence, so that the daughter of Edward, Lord Russell, the son and heir apparent of the Earl of Bedford, who had predeceased his father, could not take precedence before the wife of Sir George Carey, the son and heir apparent of Lord Hunsdon.[26] This seems to have continued to be the position until 1763, when a table of precedence prepared by John Martin Leake, Garter King of Arms, by order of the Deputy Earl Marshal assigned to the daughters of

---

[23] I.82, p. 139.
[24] I.72, p. 241. For the superseded royal warrant of the same date, see p. 35 *ante*.
[25] I.82, p. 139. As to the effect of this warrant, see p. 35 *ante*.
[26] Bodl. MS. Ashm.840, p. 127. Cooke said that this had been 'adjudged 32 Eliz.', but he did not say by whom.

the younger sons of peers precedence before the wives of the eldest sons of baronets. This was continued in a table prepared in 1812 by Sir Isaac Heard, Garter, by order of the Prince Regent.[27]

The wives of baronets have by virtue of the decree of James I made on 28 May 1612[28] precedence immediately after the wives of the younger sons of barons and the daughters of barons and before the wives of all persons before whom their husbands have precedence. By the same decree the daughters-in-law and daughters of a baronet have precedence before the daughters-in-law and daughters of knights and all other persons before whom the baronet had precedence. The daughters of a baronet have precedence next after the wife of his eldest son and before the wives of his younger sons.

The wives of knights rank in the same order as their husbands.

Apart from the comparatively rare cases of peeresses in their own right, precedence among women continued to depend entirely on that of husbands and fathers until the present century. No woman had any other right to precedence until by a royal warrant of 29 October 1912 ladies who had been or might thereafter be appointed maids of honour to Queen Mary or Queen Alexandra or any other Queen Regnant, Queen Consort, or Queen Dowager were granted the right to use the prefix of 'Honourable' with rank and precedence next after the daughters of barons.[29] However, the independent precedence of women did not become common until the institution of the Order of the British Empire in 1917.

Unlike the older orders of chivalry, the Order of the British Empire was open to women as well as men from its inception. The precedence of the members of the five classes of the new Order was defined by the statutes in the case of men by reference to that of other men and in the case of women by reference to that of other women. Thus, Knights Grand Cross were given precedence after Knights Grand Cross of the Royal Victorian Order, but Dames Grand Cross were placed before the wives of Knights Grand Cross of the Order of the Bath.[30]

---

[27] Young, *Order of Precedence*, p. 79.   [28] See p. 38 *ante*.
[29] I.75, p. 284. This warrant gave legal effect to what had been the practice during the nineteenth century (Sir Bernard Burke, *Book of Precedence* (London, 1881), p. 76). At the funeral of Elizabeth I the maids of honour were ranked after baronesses (Queen's Coll., Oxford MS. cxxii, fo.34).
[30] *Burke's Handbook to the Most Excellent Order of the British Empire* (London, 1921), p. 14.

The Order of the British Empire continued to be the only order of chivalry open to women until 1936, when the Royal Victorian Order was so opened. This made it necessary to revise the statutes of the latter Order. The revised statutes provided that women members of the Royal Victorian Order should have precedence immediately before women members of the corresponding classes of the Order of the British Empire.[31] This made sense when the statutes of the Order of the British Empire placed women members of that Order by reference to other women with settled precedence. The current statutes of that Order, however, give to women members of the various classes of the Order precedence immediately after women members of the corresponding classes of the Royal Victorian Order.[32] It has thus come about that an enquirer who relies on the statutes of the two Orders finds himself in perpetual motion as he looks from one set of statutes to the other. He can only bring himself to rest by ceasing to confine his attention to the wording of the current statutes and having regard to the history of the matter.

The statutes of the Royal Victorian Order, like those of the Order of the British Empire, are carefully drawn so as to relate the precedence of men members of the Order to that of other men and the precedence of women members to that of other women. Thus, men members of the fifth class of the Royal Victorian Order have precedence immediately after the eldest sons of knights bachelors, but women members of the same class have precedence immediately before women Members (i.e. the fifth class) of the Order of the British Empire.[33]

This recognition of the existence of separate orders of precedence for men and women was not continued when the Orders of the Bath and of St. Michael and St. George were opened to women in 1971 and 1965 respectively. In each case the precedence of both men and women members was related to the precedence of other men. Thus, Knights Grand Cross and Dames Grand Cross of the Order of the Bath have precedence next before Knights Grand Commanders of the Order of the Star of India, and Knights Grand Cross and Dames Grand Cross of the Order

---

[31] *Statutes of the Royal Victorian Order 1936*, p. 4.
[32] *Statutes of the Most Excellent Order of the British Empire 1970*, p. 12.
[33] *Statutes of the Royal Victorian Order 1936*, p. 5; *Statutes of the Most Excellent Order of the British Empire 1970*, pp. 12-13.

70 *General Precedence among Women*

of St. Michael and St. George have precedence immediately after Knights Grand Commanders of the Order of the Star of India.[34] Similarly, the precedence of the Knights Commanders and the Dames Commanders and the Companions of the Orders of the Bath and of St. Michael and St. George is related to that of the corresponding classes of the Order of the Star of India, which has no women members. This has resulted in there being no defined order of precedence between women members of the Orders of the Bath and of St. Michael and St. George and other women, but since those Orders are senior to the Royal Victorian Order and the Order of the British Empire, it must be assumed that the women members of all these Orders rank among themselves, class by class, in the order of seniority of their respective Orders.

There are thus two main types of precedence among women defined by law, namely, precedence derived from the rank of father or husband and precedence derived from membership of an order of chivalry. For completeness there may be added a third, the precedence of maids of honour under the royal warrant of 29 October 1912.[35] So far, no definite provision has been made for the precedence among other women of women holding offices formerly confined to men. An important lack of such provision was brought into prominence when a woman became Prime Minister. By the royal warrant of 20 December 1905 the Prime Minister was granted precedence next after the Archbishop of York.[36] This was clearly never intended to apply to a woman, but the effect of s.1 of the Sex Disqualification (Removal) Act 1919 was to make it potentially so applicable. It cannot be said that so to relate the precedence of a woman to that of a man is an absurdity, since that is exactly what was done by Princess Sophia's Precedence Act 1711, which placed the Electress Sophia above the Archbishop of Canterbury and has more recently been done by the exercise of the royal prerogative in the case of women members of the Orders of the Bath and of St. Michael and St. George. The royal warrant of 29 March 1972 by virtue of which Circuit judges have precedence immediately after knights bachelors[37] must also have been intended to apply to women Circuit judges, of whom

---

[34] *Statutes of the Most Honourable Order of the Bath 1972*, p. 11; *Statutes of the Most Distinguished Order of Saint Michael and Saint George 1966*, p. 6.
[35] See p. 68 *ante*.   [36] See p. 59 *ante*.
[37] See p. 56 *ante*.

there were several at that time. Thus, although women Circuit judges are ranked in their judicial capacity with their male colleagues,[38] they have no general precedence among other women. Women High Court judges are in a somewhat different position, since all those so far appointed have also been appointed Dames Commanders of the Order of the British Empire, but this does not give them precedence among women equivalent to that among men of men High Court judges, who rank well above Knights Commanders of the Order of the British Empire.

Now that some women have precedence related to that of other women, while others have precedence related to that of men, it would seem that the time has come for arranging men and women in a single order of precedence. There is, however, no way known to the law as it stands in which this can be done. Yet there are occasions when it would be highly convenient. Those who now carry out the duties which fell to the usher and marshal of a medieval nobleman's household in arranging seating places, processions and the like have to endeavour to produce a result which is in harmony with some hypothetical composite order of precedence. Not only is this legally incorrect, but it presents practical problems of which there is no authoritative solution.

---

[38] As to judicial precedence, see pp. 72-3 *post*.

CHAPTER IV

SPECIAL PRECEDENCE

Precedence can vary greatly with the occasions on which it is to be observed. Thus, while a Knight of the Garter generally yields precedence to a peer, it was provided by the statutes of the Order of the Garter made on 23 April 1522 that Knights of the Garter when wearing their mantles were to keep their places after the order of their stalls in St. George's Chapel, Windsor, and not after their estates.[1] At that time new knights succeeded to vacant stalls, so that a duke might take the stall of a knight who was not a peer or vice versa.[2] This was varied at a chapter of the Order held on 29 April 1565, when it was agreed that a new knight, other than a foreign king or prince, should be installed in the lowest stall.[3] This is still the case, so that a viscount, for example, may rank above a duke who has but recently received his blue riband. As William Segar, Norroy King of Arms, said: 'So for diuers respects the one and the other is honoured.'[4]

A similar form of special precedence is that enjoyed by a bishop within his own diocese, where he precedes all other bishops.[5]

The precedence observed in courts of law and on legal occasions, such as the services held in Westminster Abbey and Westminster Cathedral at the opening of the legal year in October, is another example of special precedence. This precedence is of considerable importance in the administration of justice, since it determines which judge shall preside over a court composed of more than one judge, and which counsel is the leader when two or more counsel are briefed together. It also regulates the order in which Queen's Counsel are entitled to move on motion days. Such precedence depends on rank in the judiciary or at the Bar. Thus a

---

[1] *Statutes of the Most Noble Order of the Garter* (London, 1766) p. 33
[2] Ibid., pp. 31-2
[3] Ibid., pp. 49-50.
[4] Segar, *Honor Military, and Ciuill,* p. 249.
[5] Coll.Arms MS. Box 40, no.42, reproduced in T. Willement, *Fac Simile of a Contemporary Roll ... of the ... Peers,* unpaginated. The following statement that the Prelate of the Order of the Garter comes next after the diocesan bishop is obsolete: see pp. 11-12 *ante.*

High Court judge or a Circuit judge who is also a peer[6] ranks with his fellow judges according to the seniority of his appointment on purely legal occasions, but as a peer on other occasions.[7] It must, however, be borne in mind that High Court judges and Circuit judges have also general precedence as such.[8] The Vice-Chancellor of the County Palatine of Lancaster has by statute judicial precedence next after High Court judges[9] and by royal warrant general precedence after knights bachelors.[10]

There are special provisions as to precedence in the Court of Appeal. The Lord Chief Justice and the Master of the Rolls when sitting and acting in that Court rank in that order, and a person who has held the office of Lord Chancellor or a lord of appeal in ordinary when so sitting and acting, ranks according to his precedence as a peer.[11]

Members of the lower judiciary, other than Circuit judges and the Vice-Chancellor of the County Palatine of Lancaster,[12] have no precedence, either general or legal, as such, though the recommendations of the Top Salaries Review Body are based on a notional ranking according to the 'weight' of the various appointments.[13]

The Attorney-General is the head of the English Bar.[14] The Attorney-General and the Solicitor-General are entitled to pre-audience in all courts of law by virtue of a royal warrant dated 14 December 1813. Before the issue of this warrant they had place and audience next after the two most senior serjeants-at-law and before the other serjeants by virtue of a royal letter issued in October 1623.[15] When appearing in the House of Lords the

---

[6] e.g. Judge Dunboyne and the late Mr Justice Finlay.

[7] A lord justice who is a peer is entitled to his precedence as a peer when sitting in the Court of Appeal: see p. 55 *ante*.

[8] See pp. 53, 55, 56 *ante*.

[9] Administration of Justice Act 1928, s.14 (1)(e).

[10] See p. 56 *ante*. He is shown in some tables of general precedence immediately after the judges of the High Court, despite the fact that he has only judicial precedence in this position.

[11] Supreme Court of Judicature (Consolidation) Act 1925, s.16(1).

[12] See p. 56 *ante*.

[13] *Review Body on Top Salaries, Report No. 10. Second Report on Top Salaries* (Cmnd.7253) (1978), p. 119.

[14] *R. v. Comptroller-General of Patents*, [1899] 1 Q.B.909, at p. 913.

[15] Young, *Order of Precedence*, p. 69.

74   *Special Precedence*

Attorney-General has precedence before the Lord Advocate, who is the leader of the Scottish Bar.[16]

During the eighteenth and early nineteenth centuries some junior barristers were granted patents of precedence, which entitled them to the precedence enjoyed by Queen's (or King's) Counsel without being subject to the disability of the latter to appear against the Crown without a special licence.[17] An early example of such a patent was that granted to Sir John Strange to take place next to the Attorney-General when he resigned his offices of Solicitor-General, King's Counsel, and Recorder of London in 1742.[18]

In the Court of Exchequer two of the most experienced barristers, called the post-man and the tub-man (from the places in which they sat) had a precedence in motions.[19]

Queen's Counsel rank in court according to the dates of their patents of appointment, and those whose patents bear the same date according to the dates when they were called to the Bar.[20] Junior barristers rank according to the dates of their call to the Bar, and those called on the same day rank according to the dates of their admission as students of an Inn of Court.

Within the Inns of Court the order of precedence is benchers, Queen's Counsel and barristers who are not benchers, and students. The Treasurer has precedence before all his fellow benchers, though many of them may rank before him in other contexts.[21] Other benchers rank according to the dates of their election to the bench, notwithstanding any personal precedence, such as knighthood, which they may have.[22]

During the Middle Ages all royal documents relating to a county were addressed to the sheriff, who was the most important man in the county during his term of office. After the appointment by Henry VIII and his successors of lords lieutenants, the sheriff,

---

[16] *Att.-Gen.* v. *Lord Advocate* (1834), 2 Cl. & F. 481.
[17] 3 Bl.Comm.28; Sir W. Holdsworth, *History of English Law* (London, 1937), vi. 476.
[18] Memorandum (1742), 2 Stra. 1176.
[19] 3 Bl.Comm. 28.
[20] Each patent states the name of the Queen's Counsel immediately after whom the appointee is to rank.
[21] e.g. I.T. Bench Table Order, 5 May 1860.
[22] e.g. I.T. Bench Table Order, 10 May 1605 (F. A. Inderwick, *Calendar of the Inner Temple Records* (London, 1896-1936), ii.10). Elias Ashmole cited this Order when advising the Secretary of State in 1664: see p. 79 *post*.

who was usually but an esquire or at best a knight, tended to be overshadowed and in some respects to be under the orders of the lord lieutenant, who was usually a peer.[23] Nevertheless, the lord lieutenant had no legal precedence over the sheriff. The only authority regarding the precedence of sheriffs was the *obiter dictum* of Coke C. J. in *Chune* v. *Pyot*,[24] where he said 'il prist le lieu de chescun noble home durant l'office'. Blackstone anglicized Coke's dictum, saying that the sheriff was the first man in the county and superior in rank to any nobleman during his office,[25] implying that the sheriff came before the lord lieutenant. However, Sir Charles Young, Garter King of Arms, challenged this interpretation, pointing out that in Coke's time the expression *noble home* did not necessarily mean a peer.[26] Young supported his point with a quotation from Coke himself, who wrote 'Nobiles sunt qui arma gentilicia antecessorum suorum proferre possunt.'[27] While Young was correct in observing that in Coke's time *noble home* was not confined to a peer, this did not prove that a sheriff during his term of office did not rank in his county before all peers. Young suggested, however, that the fact that the sheriff was mentioned in the commission of oyer and terminer, by which the judges of assize were appointed, only as an officer ministerial to the commissioners, among whom the lord lieutenant, even if a commoner, was named before the judges, indicated that the lord lieutenant had precedence before the sheriff.[28] Despite Young's criticism of Coke's *dictum,* the relative precedence of the sheriff and lord lieutenant remained in doubt. This doubt was not resolved until 19 February 1904, when a royal warrant provided that a lord lieutenant during his office and within the limits of his jurisdiction should have precedence before the sheriff having convenient jurisdiction.[29] Sheriffs appointed for a county or Greater London are now known as high sheriffs.[30]

---

[23] For lords lieutenants see G. Scott Thomson, *Lords Lieutenants in the Sixteenth Century* (London, 1923), p. 24.
[24] (1615), 1 Rolle 237.
[25] 1 Bl.Comm. 343.
[26] C.G.Y[oung], *The Lord Lieutenant and High Sheriff* (n.p.,1850), pp. 9-10.
[27] 2 Co.Inst. 594.
[28] Y[oung]. op.cit., p. 16 Commissions of oyer and terminer were abolished by the Courts Act 1971, s.1(2).
[29] I. 74, p. 45. Deputy lieutenants as such have no precedence.
[30] Local Government Act 1972, s.219(1). The sheriffs of counties had long been known colloquially as high sheriffs.

Chairmen of county councils have no precedence in that capacity outside their council chambers.

Subject to any established rule or custom affecting a particular bench, precedence among justices of the peace should be determined by seniority according to the order of the names in the commission of the peace.[31] However, the chairman or a deputy chairman of the justices in a petty sessions area has precedence at any meeting of justices for the area.[32]

The statute of apparel of 1509[33] enumerated the head officers of cities, boroughs, and corporate towns wearing apparel as their predecessors as mayors, recorders, aldermen, sheriffs, and bailiffs, no doubt following a well-established order of precedence. This was made statutory by s.57 of the Municipal Corporations Act 1835, which provided that the mayor should have precedence in all places within the borough.[34] This was replaced by s.15(5) of the Municipal Corporations Act 1882. It was also provided by s.163(5) of the Act of 1882 that the recorder should have precedence in all places within the borough next after the mayor. S.15(5) of this Act was in its turn replaced by s.18(5) of the Local Government Act 1933, subject to a proviso that nothing in that subsection should prejudicially affect the royal prerogative. S.18(5) of the Act of 1933 was replaced by s.3(4) of the Local Government Act 1972, which provides that the chairman of a district council in England shall have precedence in the district, but not so as prejudicially to affect the royal prerogative. There is a similar provision relating to Wales in s.22(4) of the Act of 1972. Many chairmen of district councils are entitled to the style of mayor by virtue of royal charters granted under s.245(1) of the Act of 1972. Such a mayor is not to be confused with a town mayor, who is the chairman of a parish council (or in Wales a community council) which has resolved under s.245(6) that the parish (or community) shall have

---

[31] Home Office Circular, dated 16 October 1907, cited in *Stone's Justices' Manual* (London, 1980), i.3.

[32] Justices of the Peace Act 1949, s.13(3). The chairman and deputy chairman have no right as such to preside at meetings of a committee or other body of justices having its own chairman, or at meetings when any stipendiary magistrate is engaged as such in administering justice (ibid., s.13(4)).

[33] 1 Hen.VIII, c.14.

[34] This conferred only social precedence and so did not entitle a mayor to preside over the borough justices: *Ex parte the Mayor of Birmingham* (1860), 3 E. & E. 222.

the status of a town.³⁵ Town mayors have no precedence *virtute officii* outside the council chamber.

The precedence of the mayor was challenged in Oxford and Cambridge by the vice-chancellor of each university. In 1612 James I appointed Lords Commissioners to hear and determine certain differences between the university and the mayor of Oxford. It was then ordered that the mayor should give precedency to the chancellor and vice-chancellor in all places within the university and city and elsewhere, although the mayor's authority was 'in his kind absolute also, and in no way subordinate to the other'.³⁶ This ruling, however, did not serve to prevent an unseemly brawl when Queen Anne visited Oxford in 1702, after which peace was made by an agreement that the university should have the right of precedence which had been awarded to it in 1612.³⁷

The controversy at Cambridge came to a head in 1647, when it was decided by the House of Lords that the vice-chancellor had precedence before the mayor.³⁸ Despite this decision, there was further trouble on 2 April 1818, when the vice-chancellor, Dr William Webb, came into court and demanded that the mayor, John Purchas, should relinquish his seat as chairman of the sessions. Purchas refused 'most positively, most unequivocally'. Webb replied that he had not come to have words and retired from the court. The corporation of Cambridge afterwards thanked Purchas for 'his firm and independent conduct in supporting the rights of his office as Mayor'.³⁹

The disagreements between the mayors and the vice-chancellors were reflected in s.137 of the Municipal Corporations Act 1835, which provided that nothing in that Act should be construed so as to alter or affect the rights or privileges of the universities of Oxford or Cambridge, and expressly recognized in s.257(2) of the Municipal Corporations Act 1882 and s.302(b) of the Local Government Act 1933, which provided that nothing in either Act should entitle the mayors of Oxford or Cambridge to any pre-

---

[35] e.g. there is a mayor of Weymouth and Portland and a town mayor of Portland.
[36] *Oxford Council Acts 1701-52* (Oxford Hist.Soc.N.S.x, 1954) 289; *Victoria County History, Oxford*, iv (Oxford, 1979), 157. The latter attributes the order to the Privy Council, but the Lords Commissioners would be those for executing the office of Earl Marshal.
[37] *Oxford Council Acts 1701-52*, pp. 13-15.
[38] 9 *L.J.* 188a.
[39] J. M. Gray, *Biographical Notes on the Mayors of Cambridge* (Cambridge, 1922), p. 56.

cedence over the respective vice-chancellors of the universities of Oxford or Cambridge respectively. There is no corresponding proviso in the Local Government Act 1972.

The recorders of boroughs ceased to exist when courts of quarter sessions were abolished by s.3 of the Courts Act 1971. A recorder of the Crown Court appointed under s.21 of that Act has no precedence by virtue of his appointment, nor has an honorary recorder appointed under s.54 of the Act.

On 31 October 1619 the Commissioners for executing the office of Earl Marshal decided that a former mayor of Oxford should have precedence before any before whom he had the priority of mayoralty, including those who were aldermen before him, and they directed that this rule be observed in all other corporations.[40] The office of alderman was abolished by the Local Government Act 1972, but it would appear that this ruling still applies to ex-mayors.

The legislation relating to local authorities in general does not apply to the City of London. The Lord Mayor of London has precedence in the City by ancient usage.[41] The Recorder of London, whose appointment was unaffected by the Courts Act 1971, is placed after the aldermen who have passed the Chair.[42] In 1607 a dispute arose as to precedency between knighted aldermen and commoners who had received the accolade before them. The aldermen petitioned the King, who referrred the matter to the Commissioners for executing the office of Earl Marshal. The knighted commoners made default of appearance, and the proceedings dragged on for four years. In the end the Commissioners decided that knighted aldermen should have precedence within the City before knighted commoners, but the Commissioners reserved the right to adjudge the contrary upon a full hearing of the cause and the proofs and allegations on both sides.[43] It does not appear that a full hearing of the cause was ever held, so that the point was still regarded as doubtful in 1664, when Sir Henry

---

[40] *Oxford Council Acts 1583-1626* (Oxford Hist.Soc. lxxxvii, 1928), p. 300. It was presumably in connection with this matter that the Commissioners consulted the judges of assize: see p. 95 *post*.

[41] *Notes in reference to the Place of the Lord Mayor in Proceedings through or within the City of London* (1862), pp. 3-31; J. Gutch, *Collectanea Curiosa* (Oxford, 1781), i.107.

[42] Miege, *Present State of Great Britain,* p. 209. For the Recorder's general precedence as a Circuit judge, see p. 56 *ante*.

[43] Gutch, *Collectanea Curiosa*, i. 99-119.

Bennet, the Secretary of State consulted Elias Ashmole, Windsor Herald and a barrister of the Middle Temple, on it. Ashmole's opinion was: 'Upon search and enquiry I finde, that in Corporations and other Societies, Precedency hath customarily been taken and allowed, according to the Seniority of their Election; and not with reference to any addition of Honour confer'd, either before or after entrance into such Corporation or Society'.[44]

It was no doubt in consequence of this opinion that when Charles II determined a dispute regarding the precedence of aldermen in the city of Bristol in the same year he ruled that 'in all places where the Body and Jurisdiction of the Citty is under any Forme', aldermen and their wives were to take place according only to their seniority as aldermen and that seniority of knighthood was not to avail them, but that 'in all other indifferent places whether within or without the Churches, in the streets or in private houses where there is no Solemne representacon of the body and Jurisdiction on the Citty' knighted aldermen were to take place according only to their seniority of knighthood and that their seniority as aldermen was not to avail them, and that the same order was to be observed between their wives, 'which is added for a finall remedy and prevention of future disputes'.[45]

Officers of Arms when acting in that capacity have an order of precedence which may not coincide with their respective ranks on other occasions. First comes Garter Principal King of Arms, followed by Clarenceux King of Arms and Norroy and Ulster King of Arms in that order. Bath King of Arms has precedence after Norroy and Ulster[46] and before the King of Arms of the Order of St. Michael and St. George,[47] who precedes the King of Arms of the Order of the British Empire.[48] After the Kings of Arms come the Heralds and Heralds Extraordinary, followed by Pursuivants and Pursuivants Extraordinary.[49] Heralds and Pursuivants rank among themselves in order of seniority, irres-

---

[44] C. H. Josten (ed.), *Elias Ashmole* (Oxford, 1966), iii. 983.
[45] Bodl.MS. Ashm. 857, p. 313.
[46] *Statutes of the Most Honourable Order of the Bath 1972*, p. 17.
[47] Ibid.; *Statutes of the Most Distinguished Order of St Michael and St George 1966*, p. 18.
[48] *Statutes of the Most Excellent Order of the British Empire 1970*, p. 22.
[49] For the precedence of officers in ordinary before extraordinary officers of the same rank, see Segar, *Honor Military and Ciuill*, p. 237.

pective of whether they have been knighted or have any other personal precedence.[50]

By a royal warrant of 18 December 1912 the Elder Brethren of Trinity House were given the style of 'Captain' and place and precedence immediately after the place and precedence which may be accorded to captains in the Royal Navy.[51] This, of course, relates only to occasions on which service rank is observed, since naval, military and air force officers have by virtue of their commissions no general precedence beyond that of esquires.[52]

[50] Josten, *Elias Ashmole,* iii. 983.
[51] I.76, fo.290.
[52] Cf. p. 43, n. 107 *ante.*

CHAPTER V

PERSONAL PRECEDENCE BY ROYAL WARRANT

The exercise of the royal prerogative is not confined to the prescription of the precedence of classes of persons or the holders of particular offices, but extends to the granting to individuals of personal precedence to which they would not otherwise be entitled. During the fifteenth and sixteenth centuries, and possibly earlier, such grants were made by letters patent under the Great Seal.[1] This procedure has long been superseded by less formal expressions of the royal pleasure. Thus, on 20 November 1619 James I declared to the Commissioners for executing the office of Earl Marshal that Sir Francis Seymour, kt., grandson of Edward, Earl of Hertford, whose elder brother had succeeded his grandfather in the earldom, should have precedence as second son of his grandfather.[2]

Since the reign of Charles I the royal prerogative has been exercised by the issue of warrants under the sign manual. These warrants are recorded in the series of volumes in the College of Arms known as the Earl Marshal's Books.[3]

The earliest of these recorded warrants is that dated 29 September 1627 whereby Sir John North, kt., and Gilbert North and Lady Coningsby, their sister, the grandchildren of Roger, Lord North, Baron of Cartlidge and children of Sir John North, whose eldest son had succeeded to the peerage, were granted precedence as the younger children of their grandfather.[4] It cannot, however, be assumed that all the warrants of this period were so recorded, for there is no record in the Earl Marshal's Books of the warrant of 31 December 1630 granting the precedence of an

---

[1] Several such patents are printed in Prynne, *Brief Animaduersions on ... the Fourth Part of the Institutes,* pp. 324-7.
[2] I.25, fo.59.
[3] Volumes 25 to 27 and 32 onwards in the 'I.Series'. The Earl Marshal's Books are also used for recording other documents: see A. R. Wagner, *Records and Collections of the College of Arms* (London, 1952), pp. 26-7.
[4] I.25, fo.59.

earl's younger son to George Talbot, brother of the Earl of Shrewsbury, who had succeeded his uncle in the earldom.[5]

The earliest royal warrants were not addressed to anyone, but later they were addressed to Garter and the other Officers of Arms and ordered to be registered in the College of Arms.[6] Finally, since the middle of the reign of Charles II the warrants have been addressed to the Earl Marshal and have directed him to cause them to be registered in the College of Arms. This he does by issuing a warrant, addressed to the kings of arms, heralds, and pursuivants, directing them to register the royal warrant in the records of the College.[7]

Unlike royal warrants relating to the precedence of particular classes of persons, e.g. the children of lords of appeal in ordinary,[8] or of the holders of particular offices, e.g. the Prime Minister,[9] which form part of the general law of precedence, warrants relating to the precedence of named persons vary the general law in order to meet the circumstances of special cases. Warrants of the latter kind, although in the same general form as warrants of general application, are more akin in nature and effect to private bill legislation. As in the case of such legislation, the procedure is initiated by a petition by the person to be benefited. The petition is addressed to the Sovereign, to whom it is submitted through the Home Office, and the warrant, if granted, is countersigned by the Home Secretary.

Such warrants have been issued for a variety of purposes. One of the earliest on record was that granted on 17 August 1603 to Penelope, Lady Rich, a daughter of Walter, Earl of Essex, whose husband was a baron, to have precedence as a descendant of the Bourchiers, Earls of Essex, presumably meaning that she was to have the precedence of an earl's daughter, which she had lost by marriage to a peer.[10] On 24 May 1663 Annabella, the wife of John Howe and illegitimate daughter of Emmanuel, Earl of

---

[5] *Shrewsbury Peerage Case* (1857), Minutes of Evidence, p. 199.

[6] e.g. I.25, fo.lllv. (29 May 1669).

[7] This procedure was first used on 5 June 1673 for the warrant granting to Mary, widow of Thomas Danby, precedence as if Col. William Ewer, her father, had survived William, Lord Ewer, her grandfather (I.25, fo.126). There are also sign manual warrants relating to precedence from 1777 to 1869 among the records of the Home Office in the Public Record Office (H.O.37).

[8] See pp. 34, 67 *ante*.   [9] See p. 59 *ante*.

[10] *C.S.P.D. 1603-10*, p. 42. For the loss of precedence by marriage, see pp. 64-6 *ante*.

Sunderland, was granted the precedence and privileges of the legitimate daughter of an earl.[11] The warrant whereby Maria Walpole of the parish of St. Margaret, Westminster was granted the same precedence as a daughter of an earl of Great Britain in 1743 was of the same character.[12] Although it is not stated in the warrant, the grantee was an illegitimate daughter of Sir Robert Walpole, who had recently been created Earl of Orford.[13] Another unusual warrant was that by which the daughters of the Archbishop of Dublin were granted the titles and precedence of daughters of a viscount of Ireland in 1678 a few years after their brother, Murrough Boyle, had been created Viscount Blessington.[14]

In recent times the most usual occasion for the presentation of such a petition has been the succession of a peer by an heir who is not his son, e.g. a grandson or a nephew. Since the father of the new peer never held the peerage, his brothers and sisters are not entitled to the precedence of the sons and daughters of a peer of his degree. This situation is normally rectified by the issue of a royal warrant granting to the brothers and sisters of the new peer the precedence, with the styles and titles, to which they would have been entitled had their father lived to succeed to the peerage.[15]

When a baronet is succeeded by an heir who is not his son, a royal warrant may be issued to give the mother of the new baronet style, rank, and precedence as a baronet's widow.[16] A similar warrant was issued in favour of the widow of the heir presumptive to a baronetcy, who had been killed on active service in 1940, and whose younger brother succeeded to the baronetcy.[17]

It has for some time been the usual practice for a warrant to be issued in favour of the widow of a man who has died after the announcement of the award of a knighthood and before receiving the accolade, granting to her the style, title, and precedence which

---

[11] *C.S.P.D. 1663-4*, p. 149. A few months earlier a warrant had been prepared for a declaration of Annabella's legitimacy on the ground that her father had been privately married to her mother (ibid., p. 61). Presumably it was decided that such a declaration was not within the royal prerogative.

[12] I.27, p. 134.

[13] *C.P.*, x.83.

[14] I.25, fo.213v.

[15] For forms of petition in such a case, see pp. 113, 114 *post*. It is not necessary to include among the petitioners a daughter married to a peer, since her precedence will depend on that of her husband: see p. 66 *ante*.

[16] e.g. Wolseley (1954), I.82, p. 56.

[17] Clifford (1958), I.82, p. 143.

would have been hers had her husband not died before his investiture.[18]

In 1912 it was announced that Sir Thomas Borthwick was to be created a baron, but he died before his letters patent had passed the Great Seal, leaving a widow and sons and daughters. The eldest son was created Baron Whitburgh, and a royal warrant was issued ordering that Lady Borthwick should enjoy the same style and title as if her husband had survived to hold the title and dignity of Baron Whitburgh, 'but without thereby conferring upon her any of the rights or privileges or the precedence belonging by statute or common law to the widow of a Peer of the Realm' and that the younger sons and the daughters should have, hold and enjoy the same rank, title, place, pre-eminence, and precedence as if their father had survived to hold the title and dignity of a baron of the United Kingdom.[19] The reason for the exclusion of the precedence of the widow of a peer was stated by Sir Alfred Scott-Gatty, Garter King of Arms, to be that because the widow of a peer derives her precedence from that of her husband, who had his precedence by virtue of the House of Lords Precedence Act 1539 according to his 'ancienty', there can be no definition under the Act of 1539 of the precedence from which the widow of a peer-designate could derive any precedence, since the peer-designate having had no letters patent of creation, had no 'ancienty'.[20]

The Borthwick precedent was followed in the case of Mrs G. V. Harding-Davies in 1980, when she was granted by royal warrant the style and title (but, as in the Borthwick case, not the precedence) to which she would have been entitled had her husband survived to be created a life peer by the name of Baron Harding-Davies.[21]

The petition for a royal warrant for precedence has to be drafted with some care. The facts which would justify the granting of the warrant should be set out in the body of the petition, and the prayer should set out the terms of the desired warrant. The allegations in the body of the petition are the basis for the drafting of

---

[18] For a form of petition in such a case, see p. 117 *post*.

[19] I.76, p. 323.

[20] If this argument is sound, it would appear to be equally applicable to the precedence of the sons and daughters of a peer-designate. I am grateful to Mr A. Colin Cole, Garter King of Arms, for drawing my attention to Garter Scott-Gatty's opinion.

[21] I.84, p. 89.

the recitals in the warrant, while the wording of the operative part of the warrant is based on the prayer of the petition. On 31 July 1924 the Deputy Earl Marshal issued an instruction that no such petition is to be prepared by an officer of arms without consultation with the Home Office.[22]

The fee on a royal warrant for precedence is £105 plus £5 for each and every beneficiary named therein. This does not apply to a royal warrant in favour of the widow of a knight-designate.[23]

[22] I.79, p. 205.
[23] Earl Marshal's warrant dated 11 March 1975 (I.83, p. 316).

CHAPTER VI
TABLES OF PRECEDENCE

Although information as to precedence is most commonly sought in the table set out in some current work of reference, no such table can be regarded as authoritative.[1] There is no such thing as 'the table of precedence'. The table in each work is no more than the editor's attempt to set out in a concise and convenient form the law on the subject. The result is no more than 'a table of precedence'.

The need for such statements of the law has been felt for many centuries by those who have had to arrange for the reception and entertainment of distinguished guests, and the need has grown as the law has become more complex. An early example of such an aid to hosts was an 'Order of All Estates' in 'an ancient Book' belonging to Crowland Abbey.[2] The wording of this 'Order' is clearly derived from the 'Order of all States of Worship and Gentry' drawn up for the coronation of Henry VI in 1429,[3] though it must have been drawn up after 1440, since it includes viscounts and their sons.

The Crowland 'Order', like the 1429 Coronation Order, sets out the precedence of laymen only. Presumably those responsible for the Abbey's hospitality were sufficiently familiar with ecclesiastical precedence not to require a similar 'Order' for churchmen. Others had need of an order embracing both ecclesiastics and laymen in arranging functions at which both were present. There were also occasions on which mixed gatherings of Englishmen and foreign visitors had to be arranged in some order of precedence.

One man who had need of a comprehensive order of precedence was John Russell, who was usher and marshal to Humphrey,

---

[1] *Pace* 14 *Halsbury's Laws of England* (4th edn.), paras. 433, 434, where the table of precedence in *Burke's Peerage* is cited as authority for the precedence of the Archbishops of Canterbury and York.

[2] Copy in Bodl. MS. Ashm. 857, p. 137.

[3] See p. 15 *ante*.

Duke of Gloucester (d.1446). Russell wrote a treatise which he called *Boke of Nurture*. Under the heading of 'The office of ussher & marshalle' he set out an order of precedence which includes the Pope, the Emperor, cardinals, and other foreigners, arranged in one series with the holders of English dignities and offices, both ecclesiastical and lay.[4] Russell says that he based his book on some earlier work.[5] What this earlier work was is not known, but it was also the source of other tables of precedence, which can be recognized by the opening words 'Pope hath no pere'. One of them was printed by Wynkyn de Worde in *The Boke of Keruynge*, published in 1513.[6]

John Russell's exemplar does not appear to have had any post-Reformation descendants. Later compilers of tables of precedence did not attempt to embrace Englishmen and aliens in a common order. Instead they worked from the 'Order' for Englishmen established by the end of the fifteenth century.[7]

Tables so produced were published by a number of authors in the seventeenth and eighteenth centuries, the earliest apparently being Thomas Milles in his *Catalogue of Honor*, published in 1610. None of these tables can be regarded as authoritative, each of them being the product of independent editorial activity by way of the addition of new and the deletion of obsolete entries. This resulted in a lack of uniformity, which weakens confidence in all of them.

This lack of uniformity had become apparent by the third quarter of the eighteenth century, when two attempts were made to produce a table of better quality. The first was by Joseph Edmondson, who cited authority for each entry, though for many he could vouch nothing beyond 'ancient usage and established custom'.[8] Although not dated, this cannot have been published before 1764, when Edmondson was created Mowbray Herald Extraordinary, for he is described on the title page as 'Mowbray Herald'. Edmondson's example was followed in 1773 by Sir William Blackstone, then a judge of the Court of Common Pleas, who included an

---

[4] Printed in 32 E.E.T.S. (1868), pp. 186-7.
[5] Ibid., p. 199.
[6] Reprinted ibid., p. 284. There is another version, entitled 'The order of going or sittyng', in Balliol Coll., Oxford MS. 354, printed ibid., p. 381.
[7] See p. 17 *ante*.
[8] J. Edmondson, *Precedency* (n.p., n.d.).

'Order of Precedence' in the fifth edition of his *Commentaries on the Laws of England*.[9] Blackstone may have been inspired by Edmondson's work, though comparison indicates that the two authors worked independently.

Blackstone's 'Order of Precedence' is the direct ancestor of the table which has appeared in successive editions of *Burke's Peerage and Baronetage*, as is shown by the retention of the eighteenth century symbols (*, †, ‖, ¶) for the notes and the lack of annotation for entries subsequently added to Blackstone's work. It has also been the ancestor of the tables of precedence appearing in numerous other publications, but since there have been many changes since Blackstone's day it has been necessary to delete some entries and to add others. Since each editor has made the deletions and additions according to his own ideas, there is again a lack of uniformity in the tables appearing in various current publications. For example, serjeants-at-law, Masters in Chancery, and Masters in Lunacy still appear in the 1979 edition of *Dod's Parliamentary Companion*, and both *Dod* and the 1977 edition of *Kelly's Handbook* include County Court judges, despite their transformation into Circuit judges by para.1(2) of Schedule 2 to the Courts Act 1971. These shortcomings are, however, venial compared with those of some other publications, one of the worst being William J. Thoms's *Book of the Court*, published in 1844, which, in spite of being dedicated by permission to Queen Victoria, contains tables of precedence of remarkable eccentricity.

The example of Edmondson and Blackstone in working from the original authorities was followed in 1851 by Sir Charles Young, Garter King of Arms, in his privately printed *Order of Precedence with Authorities and Remarks*. This, as its title implies, is not a treatise on the subject, but consists of tables of the precedence of men and women with the citation of authority for each entry. Young did not, however, confine himself to mere references, but where necessary discussed the entries at some length. Young's work seems to have been an impeccable statement of the law as it was in 1851. Having been privately printed, it had a limited circulation, and it does not seem to have been the ancestor of any of the tables in current works of reference.

---

[9] 1 Bl. Comm. 405.

Young's *Order* is now out-of-date in that some of the entries are obsolete, while there has been much subsequent legislation on the subject. Nevertheless, it is a sure foundation for the work of deletion and addition which is necessary in order to produce up-to-date tables of precedence. On it are based the tables set out in Appendix IV to this work. However, like the tables of precedence which have appeared in books of reference for many centuries, these tables have no inherent authority. They are but attempts to embody in convenient forms the existing law relating to the precedence of both men and women.

Tables of precedence cannot by themselves provide the answer to all problems of compiling a list of names or marshalling a procession in correct order, for most of the entries relate to classes of persons. Where there are two or more persons in one class to be considered, it is necessary to ascertain their relative seniority in that class. In 1881 Sir Bernard Burke, Ulster King of Arms, provided such information in his *Book of Precedence*, which set out the names of peers, baronets, knights, and companions of the various orders of chivalry in numerical order of precedence with alphabetical keys to the numbers. This seems to be the only work in which such a task has been undertaken. It contains 3,266 names, each printed twice. A modern work compiled in the same manner would contain many times the number of names. Not only are there more orders of chivalry, but in the case of the Royal Victorian Order and the Order of the British Empire there are fourth and fifth classes as well as knights (and dames) and commanders.

The relative precedence of peerages and baronetcies is readily ascertainable from the dates of creation given in current works of reference. In the case of privy councillors, knights, dames, and judges the year of appointment can be found in the current *Who's Who* and *Kelly's Handbook to the Titled, Landed and Official Classes*, but for the relative precedence of those appointed in the same year recourse must be had to the *London Gazette* for that year. The years of appointment of some companions, commanders, and members are in *Who's Who* and *Kelly's Handbook*,[10] but the only practicable way of dealing with the others is by personal enquiry.

---

[10] *Kelly's Handbook* contains all the Companions of the Orders of the Bath, the Star of India, St. Michael and St. George, and the Indian Empire and all the Commanders of the Royal Victorian Order. *Kelly* was last published in 1977 and is to be replaced by *Debrett's Handbook*.

CHAPTER VII

DISPUTES AS TO PRECEDENCE

It is said that an eighteenth-century lord mayor's ball was thrown into great confusion by a dispute for precedence between the wife of a watch-spring maker and the wife of a watch-case joint finisher.[1] But *de minimis non curat lex*. This was not the kind of dispute that Coke had in mind when he said in the section entitled 'Of Precedencie' in his *Fourth Institute,* first published in 1644, that the determination of the places and precedencies of others than lords of Parliament belonged to the Court of the Constable and Marshal, unless any question arose upon the House of Lords Precedence Act 1539, for that would be decided by common-law judges.[2]

Coke cited no authority for this proposition. He was, however, in effect, quoting himself, for as Chief Justice of the King's Bench he had presided over the court which decided *Poole and Redhead's Case*.[3] That case was not concerned with precedence, but with the fees payable by a knight on receiving the accolade. It was held that the Court of King's Bench had no jurisdiction in such a matter. However, the report contains a dictum by Coke, C. J., to the effect that the Earl Marshal has jurisdiction where there is no remedy at common law, and precedence is cited as an illustration of a matter in which he has such jurisdiction.[4]

It does not appear from either of the reports of *Ashton* v. *Jennings*[5] whether Coke's opinion or his dictum in *Poole and Redhead's Case* was cited in argument, but looking at the matter from the comtem-

---

[1] [E. Nares], *Heraldic Anomalies* (London, 1823), ii.328.

[2] 4 Co.Inst. 363. For the various styles of the Court of the Constable and Marshal, see p. 1 *ante*; pp. 91, 92 *post*.

[3] (1614), 1 Rolle 87.

[4] ' ... le marshal ad power done a luy lou le common ley ne done remedie pur precedencie, mes ceo appertain al marshall'. The common-law judges were careful not to trespass on the jurisdiction of the Court of Chivalry. In *Duke of Buckingham's Case* (1514), Keil.170, 172, they refused to answer a question as to the powers of the Lord High Constable put to them by the King on the ground that it was a matter of the King's law of arms, of which they had neither experience nor knowledge.

[5] (1674), 3 Keb. 462; 2 Lev.133.

porary common lawyer's point of view the decision cannot be faulted, for it was entirely in accordance with those authorities. It was certainly regarded as good law by Sir Matthew Hale, who was Chief Justice of the King's Bench at the time when the decision was given, for he included precedence when he dealt with the Court of the Constable and Marshal in both his *Analysis of the Law*[6] and his *History of the Common Law*[7], saying in the latter work that the Court had jurisdiction 'touching the rights of place and precedence, in cases where either acts of parliament or the king's patent, he being the fountain of honour, have not already determined it'. This passage was repeated almost verbatim by Blackstone.[8]

Neither Hale nor Blackstone cited any authority on this point, but another line of authority stemmed directly from *Poole and Redhead's Case* and *Ashton* v. *Jennings*. At the beginning of the eighteenth century Serjeant William Hawkins cited both cases in support of the statement in his *Pleas of the Crown* that '... it seems to be taken for granted in some books that disputes concerning precedency ... are proper for this court [i.e. the Court of Chivalry]'.[9] In his turn, Hawkins was cited by Matthew Bacon as authority for the statement that '... by the constant practice, and the general opinion of lawyers it [i.e. the Court of Chivalry] seems at this day to have jurisdiction concerning precedency'.[10]

Finally, Lord Goddard, Surrogate in *Manchester Corporation* v. *Manchester Palace of Varieties Ltd.*,[11] described Hawkins's opinion as 'of great value'.

Faced with such a series of authorities against him and none in his favour, some modern counsel applying to the Divisional Court of the Queen's Bench Division for an order of prohibition to restrain the Court of Chivalry from hearing a cause of precedence could hardly be surprised if his opponent were not called upon to argue. It must now be regarded as the law that the Court of Chivalry has jurisdiction in cases relating to precedence and, since that Court is a civil-law court, by necessary implication that

---

[6] (London, 1713), sect. XXXIX.
[7] (London, 1713), c.2.
[8] 3 Bl. Comm. 105.
[9] 2 Hawk. P.C., c.4. The first edition of this work was published in 1716.
[10] M. Bacon, *New Abridgement of the Law*, tit. Court of the Constable and Marshal. The first edition of this work was published in 1736.
[11] [1955] P. 133, at pp. 148-9.

the law to be applied is civil law. Nevertheless, this is one of the cases, like the doctrine of baronies by writ, in which law and history part company.[12]

An exhaustive examination of the surviving records of the Court of Chivalry has failed to disclose a single case in which precedence has been the subject of litigation in that Court.[13] It may be that the judges in *Poole and Redhead's Case* had in mind when they ascribed jurisdiction in cases of precedence to the Court of Chivalry some cases heard by the Earl Marshal (or Commissioners for executing his office) during the sixteenth and early seventeenth centuries when there was no Lord High Constable and before it was decided in 1622 that the Earl Marshal could sit in the Court of Chivalry without a Lord High Constable. This litigation is known only from secondary evidence, but it is clear that it was not conducted in a properly constituted Court of Chivalry.[14] Nevertheless, the Earl Marshal or the Commissioners were sometimes described as sitting in the Earl Marshal's Court or even the Court of Chivalry. Indeed, the evidence for the heraldic litigation of this period is contained in a manuscript entitled 'An Abstract of such causes as have received a judicial hearing in the Earle Marshall's Court before the Erles Marshalls of England or the Lords Comm$^{rs.}$ for that Office since the death of Edward Stafford Duke of Buckingham the last Constable of England'.[15] Among the cases in the 'Abstract' are several cases relating to precedence,[16] so Coke may well have thought that these cases fell within the jurisdiction of the Court of Chivalry.

It seems more likely that in these cases of precedence the Earl Marshal or the Commissioners were acting in an executive rather than a judicial capacity. As was said in an undated account of a dispute as to precedence between 'imediate' sons of barons and knighted Privy Councillors of Ireland: 'The marshalling and ordering of all Estates for place and precedency hath ever properly

---

[12] See J. H. Round, 'The Muddle of the Law', in *Peerage and Pedigree* (London, 1910), i.103-283.
[13] Squibb, *High Court of Chivalry*, p. 143. The subsequently discovered records of the Court calendared in F. W. Steer, *Catalogue of the Earl Marshal's Papers at Arundel Castle* (Harl. Soc. cxv, cxvi, 1964) have not proved any more fruitful, nor have some further uncalendared records at Arundel Castle, with abstracts of which the late Dr Steer kindly provided me.
[14] Squibb, op.cit., p. 42.   [15] Coll.Arm. MS. R. 19, fos. 179 *et seq.*
[16] Squibb, op.cit., p. 33.

and peculiarly appertained to the Constable and Marshal of England as by divers Orders and Constitutions by them set forth concerning the same may appear.'[17] Since the Constable and the Marshal had both executive and judicial functions, added to which they were sometimes specially appointed by the Crown to hear and determine particular disputes as to precedence,[18] it would not be surprising if there were to be misapprehension among lay people as to the capacity in which they were acting in a particular case. To the lawyers the distinction would be perfectly clear. For the exercise of their jurisdiction the Constable and the Marshal needed to sit as a properly constituted court following civil-law procedure, but there is no evidence in the records of their court that they ever did this in any matter relating to precedence.

The earliest disputes as to precedence of which there are records related to peers. As Coke correctly observed, these certainly did not fall within the jurisdiction of the Court of Chivalry and will be dealt with later in this chapter.[19] Disputes of the kind which Coke stated fell to be determined by the Court of Chivalry were dealt with in a variety of ways. Some were determined by the Sovereign in person. An example of this form of procedure was the dispute as to the precedence of baronets decided by James I in 1612 after hearing counsel.[20] Similarly James was petitioned by the serjeants-at-law on a question as to precedence between them and certain knights, though it does not appear that this was ever determined.[21] On 29 June 1620 James called the kings of arms and heralds before him after dinner, inquiring of York and Somerset Heralds (Ralph Brooke and Robert Treswell) upon what grounds they, with the rest of the heralds, had refused the younger sons of earls precedence before knighted Privy Councillors. York and Somerset answered so 'frivolously' that they were blamed by the King. James then proceeded to consider former lists and the written testimony of the Earls of Worcester, Nottingham, and Suffolk that the Lords Commissioners for ranking the procession from the Tower through London on 25 March 1603, after considering

---

[17] Bodl. MS. Ashm.857, p. 356; Queen's Coll., Oxford MS. CXXII, fo.31.
[18] See p. 94 *post*.
[19] See pp. 95-7 *post*.
[20] See p. 38 *ante*.
[21] *Precedence of Serjeants* (1840), 9 C. & P. 371n., at p. 372n.

former precedents with the heralds, had given the priority to the younger sons of earls, in whose favour he decided. This decision moved Sir Thomas Edmunds to use such words of the heralds as were befitting 'neither Councillor Knight or Gentleman, but a malitious depraver'.[22]

James I seems to have been particularly fond of acting the judge. By doing so in a controversy relating to land he earned a rebuke from his Chief Justice,[23] but that rebuke related only to the life, or inheritance, or goods, or fortune of his subjects. When determining a matter of precedence he was exercising the royal prerogative and was not in the technical sense of the word acting judicially. In this sphere he could say, with Justinian, 'Quod principi placuit legis habet vigorem.'[24] The Sovereign is the fountain of honour and can therefore decide questions of honour. But James I was unusual in his desire to do this in person in a quasi-judicial manner. More frequently, the Sovereign, instead of deciding such matters in person, referred them to the Earl Marshal or to the Commissioners for executing his office.

The procedure in such references was not uniform. The most formal procedure was for a petition to be presented to the Sovereign by the aggrieved party praying for the matter to be referred for the judicial hearing and determination of the Earl Marshal or the Commissioners. There was then a formal reference, and after the hearing the Earl Marshal or the Commissioners embodied their decision in an order in writing.[25] In some cases, however, the Commissioners purported to act by the authority of the commission by which they were appointed. The Commissioners who determined the precedence of 'personages of Great Estate birth and callinge' in 1595[26] stated that they did so by virtue of the commission by which they had been appointed in 1592.[27]

Similarly, on 29 March 1609 the then Commissioners decided that Sir Thomas Smith, late ambassador to the Emperor of Russia, was entitled to precedence over certain knights bachelors of the

---

[22] Bodl. MS. Ashm. 862, p. 70.
[23] *Prohibitions del Roy* (1607), 12 Co.Rep. 63.
[24] Institutes, tit.1.2.6.
[25] Gutch, *Collectanea Curiosa,* i.97 (petition and reference), 117-19 (order). For the facts of this case, see p. 78 *ante.*
[26] See pp. 24-5 *ante.*
[27] Ordinance, printed in Young, *Privy Councillors and their Precedence,* p. 49.

City of London who had been knighted before him 'by vertue of all that power and authority wch we have from his Matie by the strength of his Com to decide doubtes and questions of like nature'.[28] On these and other similar occasions the proceedings were stated to be 'at a Marshal's Court', but there is nothing to indicate that they were conducted in accordance with civil-law procedure, and the description is inconsistent with the statement that the Commissioners were acting by virtue of their commission. Such proceedings were entirely different in their nature from those of the Court of Chivalry, as is shown by the action of the Commissioners in 1619, when they referred a question regarding the precedence of ex-Mayors of Reading to the judges of assize.[29]

The Earl Marshal or the Commissioners for executing his office did not, however, have exclusive jurisdiction in cases relating to precedence. They were but referees, and it was open to the Sovereign to dispose of such cases in other ways. Charles I decided that knights of the Bath and their wives should have precedence before knights bachelors and their wives by merely directing the Earl Marshal to let his will and pleasure by known,[30] while in 1664 Charles II referred a dispute between Sir Robert Carr and Sir Robert Yeomans as to their precedence to the Privy Council.[31]

It is thus apparent that the disputes as to precedence which Coke believed to have been within the jurisdiction of the Court of Chivalry were in fact determined directly or indirectly by the Sovereign as the fountain of honour. As Charles I put it, a petition relating to precedence was directed to him as 'the source and Founteyne of all hono^r, and the supreme Judge, unto whose final decision all differences and appeals in cases of this nature, are wholly to be referred, and do principally belonge'.[32]

Questions as to the precedence of lords of Parliament, which Coke excluded from the jurisdiction of the Court of Chivalry, also fall to be determined by the exercise of the royal prerogative. However, these seem never to have been decided by the Sovereign in person but to have been the subject of references. Henry IV

[28] Bodl. MS. Ashm. 862, p. 67.
[29] *C.S.P.D. 1619-1623*, p. 17. See p. 78 *ante*.
[30] I.26, p. 25.
[31] Bodl. MS. Ashm.857, p. 415.
[32] I.25, fo.61. The petition related to the placing of the English nobility in commissions regarding the public service in Scotland and Ireland.

called a council at Westminster in 1405 to debate a dispute between the Earl of Kent and the Earl of Arundel as to their places in Parliament,[33] but the more usual course has been to refer such matters to the House of Lords. In 1449 a dispute as to precedence between the Earl of Arundel and the Earl of Devon was committed by the King and the House of Lords to the judges, who declared that it was a matter to be decided and determined by the King and the Lords Spiritual and Temporal in Parliament.[34]

The best recorded case of this kind is the dispute between John Mowbray, Earl Marshal and Richard, Earl of Warwick regarding their respective places in the Parliament of 1425. The proceedings are recorded at some length in the Rolls of Parliament.[35] Coke thought so highly of the case that he said:

He that is desirous to understand the true rules of precedency of the nobles of this realm in the high court of parliament,&c let him reade the great case between John earl Marshall and Richard Earl of Warwick, in Parliament, and the affirmations, answers, and replications on both parts exceedingly long, but full of notable rules, reasons, and presidents concerning Precedency ... very delightful to be read.[36]

Coke's praise is somewhat excessive, since the record contains nothing beyond the pleadings of the parties, each of whom was represented by counsel. No decision on the point at issue was ever given, because it was made unnecessary by the reversal of the attainder of the Earl Marshal's father, Thomas, Duke of Norfolk, which was all that stood between the Earl Marshal and the dukedom. With the attainder out of the way, his precedence before Warwick became indisputable.

By the sixteenth century it had become usual for disputes between peers as to precedence to be referred to a committee of peers for examination and report. Thus, on 31 January 1557 the Earls of Arundel and Shrewsbury and Lord Darcy of Chiche were appointed to inquire into the precedence of the ancestors of Lord Clinton and Lord Stafford.[37] The procedure was similar to that in cases relating to peerage claims, and the peers to whom such matters were referred came to be known as the Committee for Privileges.

---

[33] *Rot. Parl.* iv. 267.   [34] *Rot. Parl.* v. 148a.
[35] Ibid., iv.262b *et seqq.*, summarized in *Controversy for Precedence between John Moubray Earl Marshal and Richard Earl of Warwick* (n.p.,n.d., *c.* 1850), pp. 3-30.
[36] 4 Co. Inst. 362.   [37] 1 *L.J.* 522.

Disputes as to Precedence

The types of precedence so far discussed all have their origin in the exercise of the royal prerogative, and the determination of disputes regarding them by the further exercise of the royal prerogative can hardly be regarded as a judicial process, even though those to whom they have been referred may have heard counsel and considered evidence. The true nature of the proceedings is rather the elucidation by the legislator of his legislation. It may be otherwise with precedence which has its origin in statute. Coke stated that any question upon the House of Lords Precedence Act 1539 would have to be decided by common-law judges.[38] When Coke wrote, the Act of 1539 was the only statute relating to precedence, but there seems to be no reason for not extending what he said to the subsequent statutes conferring either general or special precedence. There appears, however, to be no reported case of a dispute as to such statutory precedence. It must also be borne in mind that, although it is provided by the Act of 1539 that peers not holding any of the offices referred to in the Act are to sit and be placed 'after their ancienty', this was merely declaratory, for it made no change in the previous law, and disputes as to 'ancienty' have not been decided by the common-law judges, but by the Committee for Privileges of the House of Lords.[39]

[38] See p. 91 *ante.*
[39] See p. 96 *ante.*

APPENDIX I

THE LORD CHAMBERLAIN'S ORDER OF 1520, AS AMENDED IN 1595[1]

On Mondaye in the Easter Weeke in the xjth yere of the raigne of King Henry the Eyght of glorious memory the Earle of Worcester then beynge Chamberlayne to the Kinge, dyned in the Greate Chamber att Richmont in his Roome and Mons. de la Batye Ambassador to the ffrenche kinge dyned with him sittinge directly on the over syde against the sayde Lord Chamberlayne, The Ambassador of Venyce, sitting next unto the L. Chamberlayne on the insyde, The Earle of Westmorland on the over syde next to the ffrench Ambassador. The Earle of Kentt on the insyde next to the Ambassador of Venyce. The Earle of Devonshire on the owtesyde next unto the Earle of Westmerlande. At whiche tyme order was taken for the placynge of Lordes and Ladyes as hereunder is sett downe.

1.—Firste the Duke to goo after his Creation, and the Duches his wyfe to goo after the same.
2 Item.—A Dukes eldiste son is borne a Marquis, savinge he shall goo beneath all Marquisses, and his wyfe beneath all Marquisses wyves, and above all Dukes daughters.
3 Item.—Dukes daughters be borne as Marquisses in all degrees, savinge they shall goo beneathe all Marquisses and Dukes eldiste sonns wyves. And yf they be married to a Baron, they shall goo after the Estate of their housbands. And if they marye with a Knight, or under the degree of a Knight, then to go after ther birth.

---

[1] The text of the Order survives in its amended form in a copy of a paper lent to Richard Lee, Clarenceux King of Arms, on 17 January 1595 by Lord Treasurer Burghley, one of the Commissioners appointed to inquire into the precedence of 'personages of great Estate birth and callinge'. A note in the handwriting of Ralph Brooke, York Herald, states that he saw Burleigh deliver the paper to Lee (Coll.Arm. MS. R.36, Hare I, p. 181). The preamble stating the circumstances in which the Order was drawn up must have been prepared after 28 January 1547, since it describes Henry VIII as 'of glorious memory'.

Appendix I: The Lord Chamberlain's Order

4 Item.—Dukes younger sonns be borne as Earles, and shall goo above all Viscounts, and beneath all the eldiste sonns of Marquisses, and ther wyves to go accordynge to the same.

5 Item.—A Marquis to goo after his Creation and the Marquisses ther Wyves to goo after the same.

6 Item.—A Marques eldiste sõne is borne an Earle and shall goo above all Dukes younger sonns and above all Viscounts and their Wyves accordinge to the same.

7 Item.—All Marquisses daughters to be borne as Countisses and shall goo above all Dukes younger sonns Wyves and above all Viscountisses, and yf they be maried to a baron they shall goo after ther housbande, and yf thay be maried to a Knight, or under the degree of a Knight, thay shall goo accordinge to ther byrthe.

8 Item.—All Marquisses younger sonns to be borne as Barons and shall goo beneath all barons and above all Viscounts eldist sonns, and ther Wyves to goo accordinge to the same.

9 Item.—An Earle to goo after his Creation and the Countisses their Wyves to goo after the same.

10 Item.—An Earles eldiste sonne is borne as a Viscounte savinge he shall goo beneath all Viscounts and his Wyfe beneath all Viscountisses and above all other Earles daughters.

11 Item.—Earles daughters are borne as Viscounts savinge thay shall goo beneath all Viscountisses and the Earles eldist sonns wyves and yf thay be maried to a baron thay shall goo after the degree of ther housbande. And yf thay marie with a Knight or under the degree of a Knight thay shall goo after theire birthe.

12 Item.—Earles younger sonnes be borne as barons sayinge thay shall goo beneath all barons and Viscounts eldiste sonns and above all Baronetts[2] and their Wyves to goo beneath all baronesses and Viscounts daughters and above all Baronetts Wyves.

13 Item.—A Viscount to goo after his Creation and the Viscountes theire wyves after the same.

14 Item.—Viscounts eldiste sonns be borne as barons and shall goo as Barons savinge thay shall goo beneath all Barons all Marquisses younger sonns and above all Earls younger sonns and their wyves shall goo beneath all baronnesses and above all Viscounts daughters.

[2] i.e. bannerets.

15 Item.—Viscounts daughters be borne as Baronesses savinge they shall goo beneath all Baronesses and Viscounts eldist sonns wyves, and yf they be maried to a Baron thay shall goo after the degree of their housbandes and yf they marye a Knighte or under the degree of a Knighte thay shall goo after theire byrthes.

16 Item.—All Viscounts younger sonns as Baronetts[3] and shall goo as Baneretts savinge thay shall goo beneath all Baneretts and theire wyves to goo accordinge to the same.

17 Item.—A Baron to goo after his Creation and the Barronesses their wyves to go after the same.

18 Item.—Barons eldiste sonns be borne as Bañerets and shall goo as Baneretts savinge they shall goo above all Baronetts[3] and all Barons younger sonns to goo above all Batchler Knights because their ffather is a Piere of the Realme.

19 Item.—All Barons daughters to goo above all Bañeretts wyves and Batchler Knightes Wyves ~~so longe as thay be unmaryed~~ and yf thay marie under degree of a Knight thay shall then goo ~~beneath~~ above all Knights wyves according to ther Birth and Estate.

_{This was set downe & ordered by the 3 Lo. Comyssioners for these purposes, 1595.}

20 Item.—Yf there be any of the degree above written come of the blood Royall or be any kynne to the Kinges highnes thay ought to stande above the degrees that they be of themselves, as a Duke above all other Dukes and so foorthe all the degrees in lyke sorte unlesse the pleasure of the Prince be to the contrarye.

---

[3] i.e. bannerets.

APPENDIX II

THE 'ANCIENTY' OF BARONS BY WRIT OF SUMMONS

It is provided by s.7 of the House of Lords Precedence Act 1539 that all peers other than office-holders with special precedence under the earlier sections of the Act shall 'sytt and be placed after ther auncients as it hathe ben accustomed'. This presents no difficulty with regard to those whose peerages were created by letters patent, but the application of this apparently simple rule to those enjoying baronies created by writ of summons has given rise to considerable difficulties and sometimes to disputes. This is because the Act requires the application of modern law to ancient facts.

Even after the composition of Parliament as the Sovereign, lords spiritual and temporal, and commons became stereotyped towards the end of the thirteenth century, the composition of what may, somewhat anarchronistically, be called the House of Lords was not equally stereotyped. The identity of those receiving personal writs of summons depended on the royal will and pleasure In some cases a man would be summoned once or twice and never again. In others he might be summoned to every parliament for the rest of his life, but his eldest son might not be summoned in his place. Gradually it became the usual practice to summon the senior male representative in each generation of the man first summoned. This, in its turn, gave rise to the belief that these heirs of the body of the man first summoned had a right to be summoned to every parliament. Finally, it came to be accepted that such a right amounted to an incorporeal hereditament, called a peerage, with the degree of a baron and the title of 'Lord'. This development was a gradual process over the fourteenth and fifteenth centuries. It can be regarded as complete by 1487, when Henry VII issued a warrant to the Lord Chancellor in the following terms: 'For as much as we have determined to call Our trusty and right well-beloved Knights and Councillors Sir John Cheyne and Sir Thomas Bourgh to the dignity and pre-eminence of Barony We therefore will and charge you that under Our Great Seal you do make out our writs according thereunto'.[1]

Again, in 1497 Richard Neville and Richard Willoughby each claimed to be entitled to the barony of Latimer. The merits of their respective claims are not material in the present context: what is important is that Neville founded his claim on the fact that William Latimer, knight, had been summoned to the parliament held in the twenty-seventh year of the reign of Edward I [1299] and had appeared and sat in that parliament as a baron and 'by that authority was later reputed and known as a baron of [the realm] by the name of Lord Latimer during his life'. This allegation was admitted by Willoughby.[2]

It may perhaps be said that the pleadings of the parties in this dispute are not conclusive of the law as it was in 1497, but the doctrine that a peerage is created by a writ of summons followed by a sitting in Parliament was enunciated by Coke,[3] and it received judicial approval in the *Clifton Peerage Case*.[4] Since then, although it has received much adverse non-judicial comment,[5] this doctrine has not been questioned in any peerage case. Its application has made it necessary to endeavour to discover at what date the creation of each barony by writ is to be deemed to have occurred, thereby determining its 'ancienty'.

It is, however, not clear how 'ancienty' was determined before the law relating to the creation of baronies by writ had developed into its modern form. There is said to be ground for supposing that certain seats in the House of Lords were formerly assigned to certain ancient baronies and that the person summoned in the name of such a barony was deemed to be entitled to the appropriate seat.[6] Some support for this is to be found in the fact that in 1513 Sir Thomas Manners, K. G. was summoned to Parliament by the style and title of Baron of Roos 'and sat in Parliament as a Peer of the Realm in the antient place of his ancestors the Lords Roos'.[7] In 1523 there was a dispute between Lord Clifford and

---

[1] *Burgh, Strabolgi, and Cobham Baronies* (1913), Minutes of Evidence and Proceedings, i.397.
[2] The whole process is printed in Powell and Wallis, *House of Lords in the Middle Ages*, App.A, pp. 583-92.
[3] Co.Litt. 16b.　　[4] (1673), Collins 291.
[5] e.g. *C.P.* ii. xiii-xv.　　[6] *C.P.* i.473.
[7] Cited in *Burgh, Strabolgi, and Cobham Baronies* (1913), Minutes of Evidence and Proceedings, i.410.

Appendix II: The 'Ancienty' of Baronies

Lord Fitzwalter 'for their Seats in the Parliament Chamber', and it was decided that Clifford should enjoy and keep 'the place and room' next above Fitzwalter.[8] The barony of De Ros was confirmed to William Cecil by letters patent dated 22 July 1616 with 'the ancient seat and place of Lord Roos in our parliaments and all other our assemblies'.[9] Similarly, in 1627 Henry, son and heir apparent of Francis, Earl of Cumberland was summoned in his father's barony of Clifford and ranked in the 'place pretended to be due to the ancient barons of Clifford'.[10]

On the other hand, it seems equally likely that these references to 'seats' and 'places' were merely figurative and that seats were arranged anew for each parliament in accordance with the 'ancienty' of the barons summoned to and attending that particular parliament. Thus at the opening of Parliament in 1529 Garter King of Arms demanded a reward for the ordering of the seats of the lords making their first entry into the parliament chamber, 'according to the Old Ordinance'.[11]

The text of the 'Old Ordinance' does not appear to have been preserved, so that the criterion of 'ancienty' required by it can only be a matter of inference. The only list of names which can be said with certainty to have been drawn up in accordance with the 'Old Ordinance' is that made by Garter in 1529, but the requirement of the Act of 1539 that peers should sit and be placed 'as it hathe ben accustomed' presumably meant that the 'Old Ordinance' should still be observed. However that may be, the names of the Lords in the Parliament Roll of 1540 can safely be regarded as having been placed 'as it hathe ben accustomed' in accordance with the Act of 1539, for the text of the Act is set out in the back of the first and second membranes of the Roll.[12] The lists of names in the surviving fragment of the 1461 journal of the

---

[8] Sir W. Dugdale, *Perfect Copy of all Summons of the Nobility* (London, 1685), p. 493. This work contains copies of some documents other than writs of summons. So far as the latter are concerned, it has been superseded by the *Report on the Dignity of a Peer* (1820), Appendix I.

[9] *Barony of Roos* (1616), Collins 162, at p. 172.

[10] 3 *L.J.* 695.

[11] Coll.Arm. MS. H.13, fo.403a, printed in Dugdale, op.cit., p. 497. The lords in question were not newly-created peers, but those sitting for the first time, some of them holding baronies by writ of considerable antiquity.

[12] Coll.Arm. MS. Box 40, No.40; *Heralds' Commemorative Exhibition 1484-1934* (London, 1936), p. 35.

House of Lords also have a strong claim to be regarded as evidence of the order required by the 'Old Ordinance' at that time.[13]

While the ranking in these documents is of considerable historical interest, it is of little, if any, practical value. Each document is confined to the baronies then represented in the House of Lords, so excluding those held by minors or by women, those dormant or in abeyance, and those subject to attainder. The later holders of many of the excluded baronies have become entitled to seats in the House of Lords, making it necessary to determine the ranking of these baronies in relation to those included in the early lists. On the other hand, some of those in the early lists have become extinct, but more importantly most of the survivors have been in abeyance at some time, some of them more than once. It has therefore been necessary to determine the ranking of each of these baronies in relation to those in actual enjoyment at the time when its abeyance has been determined.

The practice of the House of Lords in ranking the new holder of an old barony is to determine the original date of the creation of the barony partly by the facts and partly by such legal presumptions as are applicable to these facts.[14] It has been held that a barony by writ may be regarded as commencing at the date of the first summons of an ancestor to a parliament, if the facts are sufficient to enable the Committee for Privileges to infer that there must have been a sitting earlier than the first of which there is proof.[15] This has resulted in 'ancienty' considerably antedating the first recorded sitting being assigned to a number of baronies, e.g. de Clifford (65 years), Berners (54 years), Botetourt (72 years), de Ros (98 years), Zouch (49 years), Botreaux (51 years), and Mowbray (35 years).

In order to have any effect for the purpose of determining

---

[13] The fragment of the 1461 journal is printed in W. H. Dunham, *The Fane Fragment of the 1461 Lords' Journal* (New Haven, 1935). There are discrepancies between some of the lists, but they are minor, and the scribe seems to have attempted to rank the lords in order of precedence. The lists in the later journals, which begin in 1505, are less reliable, while the lists of writs of summons on the Close Rolls are not reliable at all: see N. H. Nicolas, *Report of Proceedings on the Claim to the Barony of Lisle* (London, 1829), pp. 68n., 70n., 247.

[14] Per Cozens-Hardy, K.C. *arguendo* in *Burgh, Strabolgi, and Cobham Baronies* (1913), Minutes of Evidence and Proceedings, i.381.

[15] *Burgh, Strabolgi and Cobham Baronies* (1913), Minutes of Evidence and Proceedings, i.per Lord Ashbourne, at p. 446 and per Lord Shaw, at p. 448. The matter was further discussed in *Baronies of Strange of Knokin and Stanley* (1920), Evidence and Proceedings, pp.vi-ix, and *Montacute and Monthermer Peerages Case* (1928), Proceedings and Evidence, p. xxv.

Appendix II: The 'Ancienty' of Baronies

'ancienty' the writ of summons and the subsequent sitting relied upon by the holder of the barony must both have been to an assembly which can properly be regarded as a parliament. This is a matter on which there are in some cases conflicting authorities. The modern rule, as enunciated by Lord Parker of Waddington in the *St. John Peerage Claim*,[16] is that to be a parliament an assembly must have conformed in its more essential characteristics to what is called the Model Parliament of 1295, that is to say that it must have comprised lords spiritual and temporal and elected representatives of the commons. With one exception, all subsequent assemblies have been constituted in this manner, the exception being that held at Lincoln in 1300, which was held in *Barony of Beauchamp*[17] not to have been duly and legally constituted as a parliament. Lord Parker of Waddington accepted the possibility that there might have been parliaments before 1295 so nearly resembling the Model Parliament of that year that a writ of summons to, followed by a sitting in, some or one of them would be accepted as proving the existence of a barony, but he added that each case must be considered on its merits.[18] The consideration of several such cases on their merits has resulted in some conflicting decisions.

The earliest pre-1295 assembly to have received judicial consideration was that summoned by Simon de Montfort in the name of Henry III in 1264. James I, acting on a report from the Commissioners for executing the office of Earl Marshal, held that the barony of de Ros was created by a writ of summons to, and a sitting in, this assembly.[19] This decision was followed in the *Hastings Peerage Case*,[20] where the assembly of 1264 was held to be a true parliament. This, however, was but an *obiter dictum*, for a claim based on the summons on Henry Hastings was rejected on the ground that there was no proof that he ever sat in that 'parliament'. A writ of summons to this assembly was again relied upon by Lord Stourton in his claim to the baronies of Mowbray and Segrave, but it was held not to be a true parliament

---

[16] [1915] A.C. 282, at p. 305.
[17] [1925] A.C. 153.
[18] *St. John Peerage Claim*, at p. 305.
[19] *Barony of Roos* (1616), Collins 162.
[20] (1841), 8 Cl. & F. 144.

on the ground that its proceedings were rendered null and void by c.7 of the Dictum of Kenilworth.[21]

The decision in the *Mowbray and Segrave Case* seems to have discouraged any later claimant from founding his case on Simon de Montfort's assembly of 1264, but it did not deprive the de Ros barony of the 'anciency' which it had been awarded in 1616. A decision of the House of Lords in a peerage case relates only to the particular case under consideration and is not a binding precedent in any subsequent case. Thus in *Wiltes Claim of Peerage*[22] the House of Lords refused to follow its decision given only thirty-eight years earlier in the *Devon Peerage Case*.[23] It therefore follows that a later decision does not overrule or in any way invalidate the decision in an earlier case relating to another peerage. Therefore, the de Ros barony still retains its 'anciency' of 1264, yet such a decision only stands until it is challenged in subsequent proceedings relating to the same peerage. It is the practice to make this clear by framing the decision in favour of the successful party to a dispute as to 'anciency' between two holders of baronies by writ with a saving to the other party of any right which he might in future be able to prove by better evidence.[24]

The assembly held at Shrewsbury in 1283 had long been thought not to have the status of a true parliament when it was held in 1877 that it was a parliament and that writs of summons to Roger de Mowbray and Nicholas de Segrave, followed by their sitting in that assembly, operated as the creation of baronies.[25] The 'anciency' thereby accorded to those baronies still persists, but the decision on which that 'anciency' depends was not followed in *Barony of Beauchamp*[26] because the assembly of 1283 did not contain any clergy and the representatives of the towns were summoned by separate writs and not through the sheriffs.

The 'anciency' of the barony of Hastings was fixed in 1290 by reference to the writ of summons of Henry Hastings to the assembly held in that year, which was held in the *Hastings Peerage Case*[27] to

---

[21] *Mowbray and Segrave Peerage Claims* (1887), Speeches of Counsel and Judgment, p. 65.
[22] (1869), L.R. 4 H.L. 126.
[23] (1831), 2 Dow.& Cl. 200.
[24] e.g. Lord Clifford's successful claim to precedence before Lord Fitzwalter in 1523 (Dugdale, op.cit., p. 493). Cf. pp. 102-3 *ante*.
[25] *Mowbray and Segrave Peerage Claims*, (1877) Minutes of Evidence, p. 33; Speeches of Counsel and Judgment, p. 66.
[26] [1925] A.C. 153.
[27] (1841), 8 Cl. & F. 144.

have been a true parliament. That decision still stands, but the status of the assembly of 1290 was not accepted in either *St. John Peerage Claim*[28] or *Barony of Beauchamp*.[29]

In some cases the Committee for Privileges has not fixed the date to be taken as that of the creation of a barony by writ, but has decided that the baron should rank below another barony. Thus, Vaux has been placed next below Willoughby de Broke[30] and Camoys next to Clinton.[31] Such a decision can, however, be affected by subsequent events. For example, Lord Camoys is eight places below Lord Clinton in the 1979 Parliament Roll owing to the calling of baronies out of abeyance during the nineteenth and twentieth centuries.

In order to give effect to these conflicting decisions the existing pre-1539 baronies by writ have to be ranked in the following order of precedence in accordance with the dates on which they have been deemed to have been created.[32] For this purpose baronies which have been in abeyance for more than one hundred years have not been treated as existing, since in 1927 King George V accepted the recommendation of the Select Committee on Peerages in Abeyance (the Sumner Committee) that in the absence of special circumstances or special reasons to the contrary, no abeyance should be terminated, the first commencement of which occurred more than one hundred years before the presentation of the petition.[33]

DESPENSER [24 December 1264]. Abeyance terminated by letters patent dated 25 May 1604 in favour of Dame Mary Fane with such precedence as Hugh le Despenser, sometime Justiciar of England, had enjoyed.[34] Placed above Abergavenny on 13 April 1763.[35]

DE ROS [24 December] 1264. Abeyance terminated by letters patent dated 30 July 1806 in favour of Charlotte, wife of Lord

---

[28] [1915] A.C. 282.
[29] [1925] A.C. 253.
[30] *Vaux Peerage* (1837), 3 Cl.& F. 526.
[31] *Camoys Peerage* (1839), 6 Cl. & F. 789.
[32] Where there has been no direct decision as to the date of creation, the date of the writ of summons which is consistent with the ranking of the barony has been inserted in square brackets.
[33] *Report from the Select Committee on Peerages in Abeyance* (London, 1927), p. 11.
[34] Printed in *Botreaux, Hungerford, etc. Peerage Case* (1870), Case of Countess of Loudoun, pp. 71-2.
[35] 30 *L.J.* 403.

Appendix II: The 'Ancienty' of Baronies

Henry FitzGerald, with the precedence of the ancient barony of Ros vested in Robert de Ros by writ in 49 Henry III.[36]

MOWBRAY 28 June 1283.[37] Placed next below de Ros 17 January 1878.[38]

SEGRAVE 28 June 1283.[39] Held with Mowbray.

HASTINGS [29 May] 1290.[40] Placed next below de Ros 18 May 1841.[41]

FITZWALTER 24 June 1295. Placed in 1670 as the junior baron of the reign of Edward I, although he claimed precedence before all barons then sitting as barons, alleging that it had been determined in the reign of Henry VIII that the barony should be placed next below Clifford.[42] Placed next below Hastings 3 December 1924.[43]

FURNIVALL [24 June 1295]. The resolution of the Committee for Privileges in 1912 did not fix a date for the creation of this barony, but found that it was in the reign of Edward I vested in Thomas de Furnivall, who had been summoned to the Model Parliament of 1295.[44] Placed between FitzWalter and de Clifford in the 1964 Parliament Roll. In abeyance since 1968.

GREY OF CODNOR [6 February 1299]. In abeyance since 1496, but can be regarded as an existing barony, since proceedings for the termination of the abeyance were commenced in 1926 before the report of the Sumner Committee[45] and are still depending. Here placed in accordance with the ranking in 1461, when this barony was placed above Clinton.[46]

CLINTON [6 February 1299]. Precedence fixed next above Audley by resolution of the House of Lords in 1556.[47] Placed next below FitzWalter 5 May 1965.[48]

DE CLIFFORD 29 December 1299.[49] Placed above FitzWalter 10 June 1523[50] and next above Audley 31 January 1557.[51] Abeyance terminated by letters patent dated 15 August 1734 with the

---

[36] Printed in *Botreaux, Hungerford, etc. Peerage Case* (1870), Case of Countess of Loudoun, pp. 82-6.

[37] *Mowbray and Segrave Peerage Claims* (1877), Speeches of Counsel and Judgment, pp. 66-7.
[38] 110 *L.J.* 7.
[39] *Mowbray and Segrave Peerage Claims* (1877), Speeches of Counsel and Judgment, pp. 65-6.
[40] *Hastings Peerage Case* (1841), 8 Cl. & F. 144.   [41] 73 *L.J.* 317.
[42] 12 *L.J.* 303, 361; *Fitzwalter Peerage Case* (1844), 10 Cl.& F.946,956.
[43] 157 *L.J.* 9.   [44] *Barony of Furnivall*, Minutes of Proceedings (1912), p. 32.
[45] See p. 107 ante.   [46] Dunham, *The Fane Fragment*, pp. 5, 7, 11, 14.
[47] 1 *L.J.* 522b.   [48] 197 *L.J.* 273.
[49] 14 *L.J.* 683.   [50] Coll.Arm. MS. Box 40, no.41.
[51] 1 *L.J.* 532.

Appendix II: The 'Ancienty' of Baronies

'preheminences' and 'precedencys' enjoyed by Robert, the first Lord de Clifford, who was called to Parliament by writ dated 29 December 1299.[52] Placed next above Abergavenny 24 April 1776.[53]

STRANGE OF KNOKIN [29 December] 1299.[54] Placed next below de Clifford 19 November 1963.[55] Held with the viscountcy of St David's since 1974.

BOTETOURT [13 July] 1305. Placed next after Dacre 13 April 1764.[56] Abeyance terminated by letters patent dated 23 May 1803 in favour of Henry, Duke of Beaufort with the precedence of John, Baron de Botetourt, who was summoned to Parliament as a baron in 33 Edward I.[57]

ZOUCHE OF HARRINGWORTH [16 August 1308]. Placed next below Clinton 1 February 1816.[58]

BEAUMONT [4 March 1309]. Placed next below Camoys in 1841.[59] Placed next below Zouche 4 December 1963.[60] Held with the dukedom of Norfolk since 1975.

AUDLEY [8 January 1313]. Restored after attainder by letters patent dated 3 June 1633 to James (Tuchet), Earl of Castlehaven (I.) with the 'place and precedency of George, his grandfather, formerly Baron Audley of Hely' and confirmed in 1677 by Act of Parliament with all the 'precedencies and pre-eminences thereunto belonging'.[61] Placed between Clinton and Cobham in the 1943-4 Parliament Roll.

COBHAM 8 January 1313.[62] Attainder reversed by Alexander's Restitution Act 1916. Placed between Clinton and Strabolgi in the 1917 Parliament Roll. In abeyance since 1951.

WILLOUGHBY DE ERESBY [26 July 1313]. Placed next below de Clifford 21 February 1889.[63]

---

[52] Printed in *Botreaux, Hungerford, etc. Peerage Case* (1871), Case of Countess of Loudoun, pp. 73-4.
[53] 34 *L.J.* 671.
[54] *Baronies of Strange of Knokin and Stanley* (1921). Evidence and Proceedings, p.xlix.
[55] 196 *L.J.* 20.   [56] 30 *L.J.* 572.
[57] Printed in *Botreaux, Hungerford, etc. Peerage Case* (1870), Case of Countess of Loudoun, pp. 80-1.
[58] 50 *L.J.* 419.
[59] *Beaumont Peerage* (1840), 6 Cl.& F. 268, at p. 874; 73 *L.J.* 4. It is not clear why this barony was then deemed to date only from 1432, since there was a regular succession from 1309: see *C.P.* ii.64.
[60] 196 *L.J.* 43.   [61] *C.P.* i.343; 29 & 30 Car.II, c.15 (private).
[62] *Burgh, Strabolgi, and Cobham Baronies* (1913), Minutes of Evidence of Proceedings, i. 503.   [63] 121 *L.J.* 6.

Appendix II: The 'Ancienty' of Baronies

STRABOLGI [20 October] 1318.[64] Placed next below Clinton 16 May 1916.[65]

DACRE [15 May 1321]. In 1473 Edward IV declared that Sir Richard Fiennes in right of Joan his wife was to be reputed, had, named, and called the Lord Dacre and that he was to 'keep, have and use the same seat and place in everiche of our Parliament as ... Thomas, ... late Lord Dacre had used and kept'.[66] Placed next below Strabolgi 28 May 1970.[67]

GREY DE RUTHYN 30 December 1324.[68] Placed next below Dacre with a *salvo* 10 February 1641.[69] Placed next below Zouche of Harringworth 9 February 1888.[70] In abeyance since 1963.

MALTRAVERS [23 October 1330]. Annexed to the earldom of Arundel by statute in 1627,[71] and both held with the dukedom of Norfolk since 1660.

DARCY (DE KNAYTH) 1344.[72] Placed between Grey de Ruthyn and Cromwell in the 1932 Parliament Roll.

BOTREAUX 24 February 1368.[73] Placed between Grey de Ruthyn and Camoys in the 1916 Parliament Roll. In abeyance since 1960.

CROMWELL 28 December 1375.[74] Placed next below Grey de Ruthyn 17 July 1923.[75]

CAMOYS [20 August 1383]. Held to have been created in the reign of Richard II and placed next below Clinton.[76]

BERKELEY [20 October] 1421. Abeyance terminated by letters patent dated 12 June 1893 in favour of Louisa Mary Milman with the precedence of Sir James de Berkeley in whose favour the barony was created in 1421.[77] Placed next after Clinton 10 May 1967.[78]

---

[64] *Strabolgi Barony* (1915), Minutes of Evidence and Proceedings, ii. 159. This decision is described in *C.P.* iv.750 as an 'outrage'.
[65] 148 *L.J.* 104.
[66] *C.P.* iv. 9.   [67] 202 *L.J.* 234.
[68] *Grey de Ruthyn Peerage Case* (1876) Minutes of Evidence, p. 31.   [69] 4 *L.J.* 157a.
[70] 120 *L.J.* 6.   [71] 3 Car.I., c.4 (private).
[72] *Fauconberg, Darcy (de Knayth), and Meinill Baronies* (1903), Minutes of Evidence and Proceedings, p. 217. The date appears to be based on a sitting by John Darcy, ignoring the fact that he had a writ of summons of 27 January 1332. For criticism of this decision, see *C.P.* iv. 72-3.
[73] *Botreaux, Hungerford, etc. Peerage Case* (1871), Minutes of Evidence, p. 136.
[74] *Barony of Cromwell Peerage Claim* (1922), Proceedings and Minutes of Evidence, p. lxiii. p. lxiii.
[75] 155 *L.J.* 233.
[76] *Camoys Peerage* (1839), 6 Cl. & F. 789, at p. 867; 71 *L.J.* 646; 72 *L.J.* 4.
[77] *C.P.* ii.147; vii.706.   [78] 199 *L.J.* 530.

Appendix II: The 'Ancienty' of Baronies

HUNGERFORD 7 January 1426.[79] Held with the viscountcy of St. David's since 1974.

FAUCONBERG [3 August 1429?]. The resolution of the Committee for Privileges in 1903 did not fix a date for the creation of this barony, but found only that it was an ancient barony in fee and that in the reign of Henry VI it was vested in William Nevill in right of his wife Joan.[80] In abeyance since 1948.

LATYMER 25 February 1432.[81] Placed next below Camoys 27 May 1913.[82]

DUDLEY 15 February 1440.[83] Placed next below Latymer 6 July 1916.[84]

SAYE AND SELE *Probably February 1447*. Richard Fiennes, the heir general was created Baron Saye and Sele by patent in 1603 with a proviso that he should claim no place or precedence by reason of the ancient barony, but should rank next after such other nobles as were then barons of England.[85]

DE MOLEYNS 12 February 1449.[86] Held with the viscountcy of St. David's since 1974.

BERNERS 26 May 1455. So decided 30 May 1720.[87] Placed next below Stourton (created by patent 13 May 1448) 11 May 1832.[88]

STANLEY [15 January] 1456.[89] In abeyance since 1960.

HASTINGS 26 July 1461.[90] In abeyance since 1960.

WILLOUGHBY DE BROKE [12 August 1491]. Placed between Ferrers (1461) and Eure (1544) 27 February 1696.[91]

CONYERS [17 October 1509]. Placed 'in the ancient place of the Lord Conyers' next below Stourton 3 November 1680.[92] In abeyance since 1948.

---

[79] *Botreaux, Hungerford, etc. Peerage Case* (1871), Minutes of Evidence, p. 136.
[80] *Fauconberg, Darcy (de Knayth), and Meinill Baronies* (1903), Minutes of Evidence and proceedings, p. 217.
[81] *Latymer Peerage Claim* (1912), Minutes of Proceedings and Evidence, p. 181.
[82] 145 *L.J.* 196.
[83] *Dudley Peerage Claim* (1914), Minutes of Evidence, p.xli.
[84] 148 *L.J.* 161.   [85] *C.P.* xi. 485.
[86] *Botreaux, Hungerford, etc. Peerage Case* (1871), Minutes of Evidence, p. 136.
[87] 21 *L.J.* 339.   [88] 64 *L.J.* 197.
[89] *Baronies of Strange of Knokin and Stanley* (1921), Evidence and Proceedings, p. xlix.
[90] *Botreaux, Hungerford, etc. Peerage Case* (1871), Minutes of Evidence, p. 136. This barony is not to be confused with the barony of the same name deemed to have been created in 1290: see p. 108 *ante*.
[91] 15 *L.J.* 684.   [92] 13 *L.J.* 631.

Appendix II: The 'Ancienty' of Baronies

VAUX OF HARROWDEN [29 April 1529]. Placed next below Willoughby de Broke 12 March 1838.[93]

WENTWORTH [1529]. Placed above Burgh in the 1529 Parliament.[94] Held with the earldom of Lytton since 1957.

BURGH [3 November] 1529. Sat next after Windsor.[95] Placed next below Wentworth 11 May 1880.[96] Placed next below Vaux of Harrowden 25 May 1916.[97]

BRAYE [4 December] 1529. Abeyance terminated by letters patent dated 3 October 1839 in favour of Sarah Otway Cave with the precedence of John, Baron Braye, who was summoned to Parliament by writ in 21 Henry VIII.[98]

This ranking is in accordance with the Parliament Roll of 1979 so far as regards the baronies whose holders then had seats in the House of Lords, but if there were to be a ranking *de novo* in accordance with the most recent decisions of the Committee for Privileges, it would differ from that set out above, particularly as regards the baronies deemed by earlier decisions to have been created before the Model Parliament of 1295. For example, while no dates are assigned to peerages in the Parliament Roll, it is clear that de Ros is placed first on the basis of the assumed creation of the barony in 1264. Yet if the decision in the *Mowbray and Segrave Peerage Claims*[99] were to be applied to this barony, the date would be 1299, based on the summons to William de Ros in that year.[100] That is not to say that the present ranking is wrong in law. It is not, for it follows the decisions given over the centuries with regard to particular baronies and is therefore *res judicata*. Nevertheless it demonstrates that Round was right when de described modern peerage law as a muddle,[101] and that the adverse criticism of lawyers who are not historians by historians who are not lawyers has been fully justified.

---

[93] 70 *L.J.* 153; See also *Vaux Peerage* (1837), 5 Cl. & F.526, 628.
[94] Dugdale, *Summons of the Nobility,* p. 497.
[95] *Burgh, Strabolgi, and Cobham Baronies* (1913), Minutes of Evidence and Proceedings, i. 503. This decision appears to be inconsistent with the evidence: see *C.P.* iv. 746.
[96] 148 *L.J.* 121.   [96] 112 *L.J.* 85.
[98] Printed in *Botreaux, Hungerford, etc. Peerage Case* (1870), Case of Countess of Loudoun, pp. 90-3.
[99] See pp. 105-6 *ante*.   [100] *C.P.* xi. 97.
[101] Round, 'The Muddle of the Law', in *Peerage and Pedigree* i. 103-283.

APPENDIX III
FORMS OF PETITIONS FOR PRECEDENCE

A. PETITIONS FOR ROYAL WARRANTS OF PRECEDENCE

(i) *By the Brother and Sister of a Peer who has succeeded his Grandfather*

To the Queen's Most Excellent Majesty:

> The Humble Petition of A.B. and C.D, the wife of E.D.

Sheweth:

That upon the decease of F. Earl of Wessex, which occurred on the   day of   one thousand nine hundred and   the title and dignity of Earl of Wessex devolved upon G. now Earl of Wessex as eldest son and heir of H.B. (commonly called Viscount Casterbridge) who died on or about the   day of   one thousand nine hundred and   and who whilst living was the eldest son and heir apparent of the said F. Earl of Wessex whereby according to the ordinary rules of honour Your Petitioners the brother and sister of the said G. Earl of Wessex cannot enjoy that title rank place pre-eminence and precedence which would have been due to them had their father the said H.B. (commonly called Viscount Casterbridge) survived his father the said F. Earl of Wessex and had thereby succeeded to the title and dignity of Earl of Wessex.

> YOUR PETITIONERS therefore most humbly pray that Your Majesty will be graciously pleased to ordain and declare that Your Petitioners shall have hold and enjoy the same title rank place pre-eminence and precedence as the son and daughter of an earl as would have been due to them had their father the said H.B. (commonly called Viscount Casterbridge) survived his father the said F. Earl of Wessex and had

thereby succeeded to the title and dignity of Earl of Wessex.

And Your Majesty's Petitioners will ever pray etc.

(*Signed*) A.B.
C.D.

(ii) *By the Brother and Sister of a Peer who has succeeded his Uncle*

To the QUEEN'S MOST EXCELLENT MAJESTY:

THE HUMBLE PETITION OF A.B. AND C.D. THE WIFE OF E.D.

SHEWETH:

That upon the decease of F. Viscount Barchester which occurred on the   day of   one thousand nine hundred and   the title and dignity of Viscount Barchester devolved upon G. now Viscount Barchester, as eldest son and heir of H.B. (commonly called the Honourable H.B.) who died on or about the   day of   one thousand nine hundred and   and who whilst living was the next brother and heir presumptive of the said F. Viscount Barchester whereby according to the ordinary rules of honour Your Petitioners the brother and sister of the said G. Viscount Barchester cannot enjoy that title rank place pre-eminence and precedence which would have been due to them had their father the said H.B. (commonly called the Honourable H.B.) survived his brother the said F. Viscount Barchester and had thereby succeeded to the title and dignity of Viscount Barchester.

YOUR PETITIONERS therefore most humbly pray that Your Majesty will be graciously pleased to ordain and declare that Your Petitioners shall have hold and enjoy the same title rank place pre-eminence and precedence as the son and daughter of a viscount as would have been due to them had their father the said H.B. (commonly called the Honourable H.B.) survived his brother the said F. Viscount Barchester and had thereby succeeded to the title and dignity of Viscount Barchester.

And Your Majesty's Petitioners will ever pray etc.

  (*Signed*) A.B.
       C.D.

(iii) *By the Widow and Children of a Life Peer-designate*

To the QUEEN'S MOST EXCELLENT MAJESTY:

  THE HUMBLE PETITION OF A.B. WIDOW OF C.B. DECEASED AND D.B. AND E. WIFE OF F.G.

SHEWETH:

1. That in [*month*] one thousand nine hundred and   Your Majesty was graciously pleased to signify Your Majesty's Royal Intention to confer upon C.B. the dignity of a Peer for Life
2. That the said C.B. departed this life upon the   day of one thousand nine hundred and   before Your Majesty's Letters Patent of Creation had passed Your Majesty's Great Seal leaving Your Petitioner A.B. his relict, D.B. his son, and E. wife of F.G. his daughter him surviving
3. By reason of the decease of the said C.B. before Your Majesty's Royal Intention could be carried into effect Your Petitioners cannot enjoy the rank title place pre-eminence and precedence which would have been theirs had the said C.B. survived and received from Your Majesty the title and dignity of a Baron for Life of Your Majesty's United Kingdom.

  YOUR PETITIONERS therefore most humbly pray that Your Majesty will be graciously pleased to take into Your Royal Consideration the circumstances aforesaid and to ordain and declare that Your Petitioner the said A.B. shall have hold and enjoy the same style and title to which she would have been entitled had her husband the said C.B. survived and received from Your Majesty the title and dignity of a Baron for Life of Your Majesty's United Kingdom and that the said D.B. and E. wife of F.G. as children of the said C.B. shall have hold and enjoy the same rank title place pre-eminence and precedence as if their said father had survived and received from Your Majesty the title and dignity of a Baron for Life of Your Majesty's United Kingdom.

And Your Majesty's Petitioners will ever pray etc.

>   (*Signed*)   A.B.
>   D.B.
>   E.G.

(iv) *By the Widow of the Heir Apparent of a Baronet*

To the QUEEN'S MOST EXCELLENT MAJESTY:

THE HUMBLE PETITION OF A.B. WIDOW OF C.B. DECEASED

SHEWETH:

That upon the decease of Sir D.B. which occurred on the   day of   one thousand nine hundred and   the dignity of a Baronet devolved upon E.B. as son and heir of C.B. esquire who departed this life on the   day of   one thousand nine hundred and   and who while living was the son and heir apparent of the said Sir D.B. whereby according to the ordinary rules of honour Your Petitioner cannot enjoy that style title place and precedence to which she would have been entitled had her husband survived his father and thereby succeeded to the title and dignity of a Baronet.

> YOUR PETITIONER therefore most humbly prays that Your Majesty will be graciously pleased to ordain and declare that Your Petitioner shall henceforth have hold and enjoy the same style title place and precedence to which she would have been entitled had her said husband succeeded to the title and dignity of a Baronet.

And Your Majesty's Petitioner will ever pray etc.

>   (*Signed*)   A.B.

(v) *By the Widow of a Knight Bachelor-designate*

To the QUEEN'S MOST EXCELLENT MAJESTY:

>   THE HUMBLE PETITION OF A.B.

SHEWETH:

1. That it was Your Majesty's Royal Intention to have conferred the title and dignity of a Knight Bachelor upon Your Petitioner's late husband C.B. deceased

2. That by reason of the decease of the said C.B. before Your Majesty's Royal Intention could be finally carried into effect Your Petitioner cannot enjoy that style title rank and precedence to which she would have been entitled had her said husband survived and received from Your Majesty the Honour of Knighthood.

>   YOUR PETITIONER therefore most humbly prays that Your Majesty will be graciously pleased to take into Your Royal Consideration the circumstances aforesaid and to ordain and declare that Your Petitioner shall have hold and enjoy the same style title rank and precedence to which she would have been entitled had her husband the said C.B. survived and received from Your Majesty the degree style and title of a Knight Bachelor of Your Majesty's Realms.

And Your Majesty's Petitioner will ever pray etc.

>   (*Signed*)   A.B.

Appendix III: Forms of Petitions

B. PETITION BY A BARON BY WRIT FOR PRECEDENCE IN PARLIAMENT

To the RIGHT HONOURABLE the LORDS SPIRITUAL and TEMPORAL in PARLIAMENT ASSEMBLED:

THE HUMBLE PETITION OF N. LORD A.

SHEWETH:

1. That Your Petitioner's ancestor M.A. was created a Lord of Parliament by writ of summons to Parliament dated on the   day of   in the   year of the reign of King   and was frequently present in Parliament
2. That the first Lord A. died in the year   and was succeeded by his   as second Lord A.
3. That [*set out devolution of the peerage*]
4. That Your Petitioner succeeded as Lord A. on the   day of   19   and subsequently took his seat in Parliament when he was placed after the Lords X. Y. and Z.
5. That Your Petitioner conceives and submits that he is not ranked in his due and proper place as a Lord of Parliament.

> YOUR PETITIONER therefore most humbly prays that he may be restored to and placed in his proper precedency in Your Lordship's House and that he may henceforth be ranked next after the Lord W.; and that Orders may be accordingly issued for so placing Your Petitioner on the Roll of the Lords Spiritual and Temporal.

And Your Lordships' Petitioner will ever pray etc.

(*Signed*)   A.

APPENDIX IV
MODERN TABLES OF PRECEDENCE

(i) MEN

The Queen[1]
Prince Philip, Duke of Edinburgh
The Prince of Wales
The Queen's Younger Sons
Dukes of the Blood Royal[2]
Prince Michael of Kent[3]
Vicegerent in Spirituals[4]
Archbishop of Canterbury
Lord Chancellor[5]
Archbishop of York
Prime Minister
Lord High Treasurer[6]
Lord President of the Council
Speaker of the House of Commons
Lord Privy Seal
Ambassadors and High Commissioners in order of seniority based on dates of arrival in the United Kingdom
Lord Great Chamberlain[7]   )
Lord High Constable[8]   ) Above all
Earl Marshal   ) peers of
Lord High Admiral[9]   ) their own
) degree
Lord Steward of the Household   )
Lord Chamberlain   )
Master of the Horse[10]

---

[1] See p. 62 *ante*.
[2] Any successors of the present dukes of the blood royal will rank above all other dukes in accordance with the Order of 1520, but below the Archbishops and the Great Officers of State: see p. 20 *ante*.
[3] See p. 29 *ante*.   [4] Office vacant since 1540.
[5] Or Lord Keeper, if a peer.   [6] Office in commission since 1714.
[7] Office executed by deputy: see p. 47 *ante*.
[8] No permanent holder of this office since 1521.
[9] Office in commission since 1828.   [10] See p. 60 *ante*.

Dukes of England
Dukes of Scotland
Dukes of Great Britain
Dukes of Ireland created before 1801[11]
Dukes of the United Kingdom and Dukes of Ireland created after 1800[12]
Eldest sons of Dukes of the Blood Royal
Marquesses of England
Marquesses of Scotland
Marquesses of Great Britain
Marquesses of Ireland created before 1801
Marquesses of the United Kingdom and Marquesses of Ireland created after 1800
Eldest sons of Dukes not of the Blood Royal
Earls of England
Earls of Scotland
Earls of Great Britain
Earls of Ireland created before 1801
Earls of the United Kingdom and Earls of Ireland created after 1800
Younger sons of Dukes of the Blood Royal
Eldest sons of Marquesses
Younger sons of Dukes not of the Blood Royal
Viscounts of England
Viscounts of Scotland
Viscounts of Great Britain
Viscounts of Ireland created before 1801
Viscounts of the United Kingdom and Viscounts of Ireland created after 1800
Eldest sons of Earls
Younger sons of Marquesses
Bishop of London
Bishop of Durham
Bishop of Winchester
Other English Diocesan Bishops according to their seniority of consecration
Suffragan and retired Bishops
Secretary of State, if a baron

[11] The Duke of Leinster is now (1981) the only such Duke.
[12] The Duke of Abercorn is now (1981) the only Duke of Ireland created since 1800.

Barons of England
Barons of Scotland
Barons of Great Britain
Barons of Ireland created before 1801
Barons of the United Kingdom, Barons of Ireland created since 1800, Lords of Appeal in Ordinary and Life Peers according to their dates of appointment or creation
Commissioners of the Great Seal[13]
Treasurer of the Household
Comptroller of the Household
Vice-Chamberlain of the Household
Secretary of State, if under the degree of a baron
Eldest sons of Viscounts
Younger sons of Earls
Eldest sons of Barons
Knights of the Garter
Knights of the Thistle
Knights of St Patrick
Privy Councillors
Chancellor of the Order of the Garter[14]
Chancellor of the Exchequer
Chancellor of the Duchy of Lancaster
Lord Chief Justice of England
Master of the Rolls
President of the Family Division of the High Court
Lords Justices of Appeal[15]
Judges of the High Court in order of appointment, irrespective of the Divisions to which they are assigned
Younger sons of Viscounts
Younger sons of Barons and sons of Lords of Appeal in Ordinary, Life Peers and Life Peeresses
Baronets
Knights Grand Cross of the Order of the Bath

---

[13] See p. 61 *ante*. The Great Seal has not been in commission, except for very short periods, since 1850.

[14] So placed by a decree of the Chapter of the Order on 23 April 1629, but since the Chancellorship was annexed to the bishopric of Salisbury in 1669 and transferred to the bishopric of Oxford in 1837, this ranking has ceased to be effective.

[15] The Lord Chief Justice, the Master of the Rolls, the President of the Family Division, and the Lords Justices are usually appointed Privy Councillors and so have precedence as such, if not also peers.

Knights Grand Commanders of the Order of the Star of India
Knights Grand Cross of the Order of St Michael and St George
Knights Grand Commanders of the Order of the Indian Empire
Knights Grand Cross of the Royal Victorian Order
Knights Grand Cross of the Order of the British Empire
Knights Commanders of the Order of the Bath
Knights Commanders of the Order of the Star of India
Knights Commanders of the Order of St Michael and St George
Knights Commanders of the Order of the Indian Empire
Knights Commanders of the Royal Victorian Order
Knights Commanders of the Order of the British Empire
Knights Bachelors
Vice-Chancellor of the County Palatine of Lancaster
Recorder of London
Recorders of Liverpool and Manchester according to priority of appointment
Common Serjeant
Other Circuit Judges according to the priority or order of their respective appointments[16]
Master of the Court of Protection[17]
Companions of the Order of the Bath
Companions of the Order of the Star of India
Companions of the Order of St Michael and St George
Companions of the Order of the Indian Empire
Commanders of the Royal Victorian Order
Commanders of the Order of the British Empire
Companions of the Distinguished Service Order
Members of the Royal Victorian Order (4th class)
Officers of the Order of the British Empire
Companions of the Imperial Service Order
Eldest sons of the younger sons of Peers
Eldest sons of Baronets
Eldest sons of Knights
Members of the Royal Victorian Order (5th class)
Members of the Order of the British Empire
Younger sons of Baronets
Younger sons of Knights

---

[16] For the precedence of some Circuit Judges by virtue of appointments held before 1 January 1972, see p. 57 *ante*.

[17] Sed qu. See pp. 57-8 *ante*.

Appendix IV: Modern Tables of Precedence

(ii) WOMEN

The Queen
Queen Elizabeth, The Queen Mother
The Queen's Daughter
The Queen's Sister
Wives of Dukes of the Blood Royal[18]
Princess Alexandra of Kent, the Hon. Mrs Angus Ogilvy[19]
Duchesses of England
Duchesses of Scotland
Duchesses of Great Britain
Duchess of Leinster[20]
Duchess of the United Kingdom and the Duchess of Abercorn[21]
Wives of the eldest sons of Dukes of the Blood Royal[22]
Daughters of Dukes of the Blood Royal
Marchionesses of England
Marchionesses of Scotland
Marchionesses of Great Britain
Marchionesses of Ireland created before 1801
Marchionesses of the United Kingdom and Marchionesses of Ireland created after 1800
Wives of the eldest sons of Dukes not of the Blood Royal
Daughters of Dukes not of the Blood Royal not married to Peers
Countesses of England
Countesses of Scotland
Countesses of Great Britain
Countesses of Ireland created before 1801
Countesses of the United Kingdom and Countesses of Ireland created after 1800
Wives of the younger sons of Dukes of the Blood Royal[22]
Wives of the eldest sons of Marquesses
Daughters of Marquesses not married to Peers
Wives of the younger sons of Dukes not of the Blood Royal
Viscountesses of England
Viscountesses of Scotland

---

[18] See p. 119, n. 2 *ante*.
[19] Cf. p. 29 *ante*.
[20] See p. 120 *ante*.
[21] See pp. 31-2 *ante*.
[22] None in 1981.

Viscountesses of Great Britain
Viscountesses of Ireland created before 1801
Viscountesses of the United Kingdom and Viscountesses of Ireland created after 1800
Wives of the eldest sons of Earls
Daughters of Earls not married to Peers
Wives of the younger sons of Marquesses
Baronesses of England
Baronesses of Scotland
Baronesses of Great Britain
Baronesses of Ireland created before 1801
Baronesses of the United Kingdom, Baronesses of Ireland created after 1800, Life Peeresses, Wives of Lords of Appeal in Ordinary, and Wives of Life Peers.
Wives of the eldest sons of Viscounts
Daughters of Viscounts not married to Peers
Wives of the younger sons of Earls
Wives of the eldest sons of Barons
Daughters of Barons, Lords of Appeal in Ordinary, Life Peers, and Life Peeresses not married to Peers
Maids of Honour
Wives of Knights of the Garter
Wives of the younger sons of Viscounts
Wives of the younger sons of Barons
Wives of Baronets
Dames Grand Cross of the Order of the Bath
Dames Grand Cross of the Order of St Michael and St George
Dames Grand Cross of the Royal Victorian Order
Dames Grand Cross of the Order of the British Empire
Wives of Knights Grand Cross of the Order of the Bath
Wives of Knights Grand Commanders of the Order of the Star of India
Wives of Knights Grand Cross of the Order of St Michael and St George
Wives of Knights Grand Commanders of the Order of the Indian Empire
Wives of Knights Grand Cross of the Royal Victorian Order
Wives of Knights Grand Cross of the Order of the British Empire
Dames Commanders of the Order of the Bath
Dames Commanders of the Order of St Michael and St George

Dames Commanders of the Royal Victorian Order
Dames Commanders of the Order of the British Empire
Wives of Knights Commanders of the Order of the Bath
Wives of Knights Commanders of the Order of the Star of India
Wives of Knights Commanders of the Order of St Michael and
    St George
Wives of Knights Commanders of the Order of the Indian Empire
Wives of Knights Commanders of the Royal Victorian Order
Wives of Knights Commanders of the Order of the British Empire
Wives of Knights Bachelors
Companions of the Order of the Bath
Companions of the Order of St Michael and St George
Commanders of the Royal Victorian Order
Commanders of the Order of the British Empire
Wives of Companions and Commanders of the Orders of the
    Bath, the Star of India, St Michael and St George, and the
    Indian Empire, the Royal Victorian Order, and the British
    Empire
Members of the Royal Victorian Order (4th class)
Officers of the Order of the British Empire
Wives of Companions of the Distinguished Service Order
Wives of Members of the Royal Victorian Order (4th class)
Wives of Officers of the Order of the British Empire
Companions of the Imperial Service Order
Wives of Companions of the Imperial Service Order
Wives of the eldest sons of the younger sons of Peers
Daughters of the younger sons of Peers
Wives of the eldest sons of Baronets
Daughters of Baronets
Wives of the eldest sons of Knights
Daughters of Knights
Members of the Royal Victorian Order (5th class)
Members of the Order of the British Empire
Wives of Members of the Royal Victorian Order (5th class)
Wives of Members of the Order of the British Empire
Wives of the younger sons of Baronets
Wives of the younger sons of Knights

BIBLIOGRAPHY

I MANUSCRIPTS

1. *British Library*

Cotton MS. Faustina C. VIII   Harleian MS. 642

2. *College of Arms*

Box 40, Nos. 40-2   H.13
Chapter Book   R.19
Earl Marshal's Books   R.36
                         Vincent 151

3. *Inner Temple*

Petyt MS. 538/44

4. *Oxford, Bodleian Library*

Ashmole MS. 840   Rawlinson MS. D. 766
Ashmole MS. 857
Ashmole MS. 862

5. *Oxford, Balliol College*

MS. 354

6. *Oxford, The Queen's College*

MS. CXXII   MS. CXXXIII

7. *Public Record Office*

Domesday Book (E 31)   Sign Manual Warrants (HO 37)
State Papers James I (SP 14)

II PRINTED BOOKS

(Law reports, works of general reference, and books mentioned incidentally are not included. Unless otherwise indicated books are published in London.)

Anstis, J., *Observations Introductory to an Historical Essay, upon the Knighthood of the Bath*, 1725.

Ashmole, E., *The Institution, Laws and Ceremonies of the Most Noble Order of the Garter*, 1672.
Bacon, M., *New Abridgement of the Law*, 1736
Beaver, J., *History of the Roman or Civil Law*, 1724.
Blackstone, W., *Commentaries on the Laws of England* (7th edn.), Oxford, 1778.
The Boke of Keruinge (32 E.E.T.S.), 1868.
Bowyer, G., *Commentary on the Constitutional Law of England*, 1841.
Burke, B., *Book of Precedence*, 1881.
Burke's Handbook to the Most Excellent Order of the British Empire, 1921.
Calendar of State Papers, Domestic, 1856-.
Camden, W., *Britannia*, 1772.
C[okayne], G.E., *Complete Baronetage*, 1900-9.
   Complete Peerage (2nd edn.), 1910-59.
Coke, E., *Institutes of the Laws of England*, 1628-44.[1]
Collins, A., *Proceedings, Precedents and Arguments on Claims and Controversies concerning Baronies by Writ, and other Honours*, 1734.
Controversy for Precedence between John Moubray Earl Marshal and Richard Earl of Warwick, n.p., n.d., c.1850.
The Decree and Establishment of the King's Majestie vpon a Controuersie of Precedence, betweene yonger sonnes of Viscuntes and Barons and the Baronets, 1612.
Duck, A., *De Usu et Authoritate Juris Civilis*, Leyden, 1654.
Dugdale, W., *A Perfect Copy of all Summons of the Nobility to the Great Councils and Parliaments of this Realm*, 1685.
Dunham, W.H., *The Fane Fragment of the 1461 Lords' Journal*, New Haven, 1935.
Edmondson, J., *Precedency*, n.p., n.d.
[Egmont, John (Percival), Earl of], *The Question of the Precedency of the Peers of Ireland in England, fairly stated*, Dublin, 1739.
Finett, J., *Finetti Philoxensis; Or some choice Observations of Sir John Finett, Knight, and Master of the Ceremonies to the two last Kings, touching the Reception and Precedence, the Treatment and Audience, the Puntillios and Contests of Forren Ambassadors in England*, 1656.
Galbraith, V.H., *The Making of Domesday Book*, Oxford, 1961.
Gray, J.M., *Biographical Notes on the Mayors of Cambridge*, Cambridge, 1922.

---

[1] The first part of the *Institutes* is Coke's commentary on Littleton's *Tenures* and is cited as 'Co. Litt.'

Bibliography

Greville, C. C. F., *The Greville Memoirs (Second Part)*, 1885.
    The Precedence Question. 1840.
Gutch, J., *Collecteana Curiosa*, Oxford, 1781.
Hale, M., *Analysis of the Law*, 1713.
    History of the Common Law, 1713.
Hawkins, W., *Pleas of the Crown*, 1716.
Heralds' Commemorative Exhibition 1484-1934, 1936.
Holdsworth, W., *History of English Law* (2nd edn.), 1922-52.
Inderwick, F. A., *Calendar of the Inner Temple Records*, 1896-1937.
Johnson, C. (ed.), *Dialogus de Scaccario*, 1950.
Josten, C. H. (ed.), *Elias Ashmole*, Oxford, 1966.
Journals of the House of Lords, n.p., n.d.
Jurisprudentia Heroica, Brussels, 1668.
Luttrell, N., *Brief Historical Relation of State Affairs*, Oxford, 1857.
Macdonald, A. J., *Lanfranc*, Oxford, 1926.
Meige, G., *The Present State of Great Britain*, 1707.
Milles, T., *Catalogue of Honor*, 1610.
Myers, A. R. (ed.), *The Household of Edward IV*, Manchester, 1959.
Nicolas, N. H., *Observations on the Clauses containing Grants of Precedency in Patents of Peerage: with Remarks on the Statute 31 Hen. VIII for placing the Lords*, [1832?].
    Report of Proceedings on the Claim to the Barony of Lisle, 1829.
    Report of Proceedings on the Claim to the Earldom of Devon, 1832.
Notes in reference to the Place of the Lord Mayor in Proceedings through or within the City of London, n.p., 1852.
Oxford Council Acts 1583-1626 (Oxford Hist.Soc. lxxxvii), 1928.
Oxford Council Acts 1701-52 (Oxford Hist.Soc. N.S.x.), 1954.
Perkins, J., *The Most Honourable Order of the Bath*, 1913.
Pixley, F. W., *History of the Baronetage*, 1900.
Powell, J. E. and Wallis, K., *The House of Lords in the Middle Ages*, 1968.
Prynne, W., *Brief Animaduersions on Amendments of and Additional Explanatory Records to the Fourth Part of the Institutes of the Lawes of England*, 1669.
Regesta Regum Anglo-Normannorum, Oxford, 1913.
Robinson, J. Armitage, *Somerset Historical Essays*, 1921.
Rotuli Parliamentorum, 1783.
Round, J. H., *The King's Serjeants and Officers of State*, 1911.
    'The Muddle of the Law', in *Peerage and Pedigree*, 1910.
Russell, J., *The Boke of Nurture* (32 E.E.T.S.), 1868.

Sandford, F., *Genealogical History of the Kings and Queens of England*, 1707.
Segar, W., *Honor Military and Ciuill*, 1602.
Sharpe, K., *Sir Robert Cotton*, Oxford, 1979.
Squibb, G. D., *The High Court of Chivalry*, Oxford, 1959.
State Papers during the Reign of Henry the Eighth, 1830-2.
Statutes of the Imperial Service Order 1970.
Statutes of the Most Distinguished Order of Saint Michael and Saint George 1966.
Statutes of the Most Excellent Order of the British Empire 1970.
Statutes of the Most Honourable Order of the Bath 1972.
Statutes of the Most Noble Order of the Garter, 1766.
Statutes of the Order of Companions of Honour 1919.
Statutes of the Royal Victorian Order 1936.
Stephens, W., Mrs. *Abigail, or an Account of a Female Skirmish between the Wife of a Country Squire, and the Wife of a Doctor in Divinity*, 1700.
Stone's Justices' Manual, 1980.
Stubbs, W., *Select Charters* (9th edn.). Oxford, 1913.
[Swift, J.], *The Right of Precedence between Phisicians (sic) and Civilians enquired into*, Dublin, 1720.
Thompson, A. Hamilton, *English Clergy in the Later Middle Ages*, Oxford, 1947.
Truro, Thomas (Wilde), Lord, *On the Precedency of Peers of the United Kingdom*, c.1850.
Victoria County History, Oxford, iv. Oxford, 1979.
Wagner, A. R., *Heralds of England*, 1967.
    Heralds and Heraldry. (2nd. edn.). Oxford, 1956.
Wagner, A. and Squibb, G. D., 'Precedence and Courtesy Titles', in 89 *Law Quarterly Review*, 1973.
Whitelock, D., *English Historical Documents* (2nd. edn.), 1979.
Wilkins, D., *Concilia Magnae Britanniae*, 1737.
Willement, T., *Fac Simile of a Contemporary Roll with the Names and the Arms of the Sovereign and of the Spiritual and Temporal Peers who sat in Parliament held at Westminster on the 5th February in the Sixth year of the reign of King Henry the Eighth, Anno Domini 1515*, 1829.
William of Malmesbury, *De Gestis Pontificum Anglorum* (52 Rolls Series), 1870.
William of Newburgh, *Historia Rerum Anglicarum* (82 Rolls Series), 1884.

[Young, C. G.], *Ancient Tables of Precedency*, n.p., n.d.
[Young, C. G. ?] *Serjeants at Law* [1864 ?].
Y[oung], C. G., *The Lord Lieutenant and High Sheriff*, n.p. 1850.
Young, C. G., *Order of Precedence*, n.p. 1851.
   Privy Councillors and their Precedence, n.p., 1860.
Zouche, R., *Elementa Jurisprudentiae*, Amsterdam, 1652.

INDEX

Abbesses, 63
Abbots, 7, 8, 9, 12, 21, 48: *see also* Glastonbury, Westminster
Abercorn, Duchess of, 123
Acland, baronetcy, 38
Adams, George Edward, Rouge Dragon Pursuivant, 5
Albert of Saxe-Coburg and Gotha, Prince, 27
Aldermen, 76, 78; knighted, 78-9: *see also* Bristol
Alexandra of Kent, Princess, 123
Ambassadors, 48, 119; French, 98; Venetian, 98
Ancaster, George Henry (Heathcote-Drummond-Willoughby), Earl of, 47; Robert (Bertie), Duke of, 46-7
Ancienty, 30, 97, 101-2
Anglo-Norman precedence, 9
Anglo-Saxon precedence, 7
Anne, Queen, 77
Anselm, abp. of Canterbury, 10
Appeal, Court of, 55, 73; *ex officio* judges of, 55; vice-president of, 55
Appeal in Chancery, Court of, 55
Archbishops, 7, 8, 9-10, 12, 20, 21, 48, 49; Irish, 50: *see also* Armagh, Canterbury, Dublin, York
Archdeacons, 21
Armagh, abp. of, 49
Arundel, John (FitzAlan), Earl of, 96; Thomas (FitzAlan), Earl of, 96
Ashmole, Elias, Windsor Herald, 74, 79
Ashton, Margaret, 1, 62-3; Robert, 1
Attorney-General, 73-4
Audley, barony, 109

Bacon, Sir James, Vice-Chancellor, 54; Matthew, 91
Bailiffs, 9, 76
Banbury, William (Knollys), Earl of, 31
Bankruptcy, Court of, 54; Court of Review in, 54
Baronesses, 100, 124
Baronets, 36-40, 93, 121
  daughters of, 37, 68, 125
  daughters-in-law of, 37, 68
  eldest sons of, 122; wives of, 125
  patents of, 37, 38, 39-40
  sons of, 37, 39-40, 44
  widows of, 65
  wives of, 37, 68, 124
  younger sons of, 122; wives of, 125
  see also England, Great Britain, Ireland, Nova Scotia, United Kingdom
Barons, 8, 9, 12, 13, 21, 22, 30, 49, 100, 121
  daughters of, 100, 124
  eldest sons of, 34, 100, 121; wives of, 124
  sons of, 15
  younger sons of, 37, 38, 100, 121; wives of, 124
Barons of the Exchequer, 19, 38, 63; cursitor, 53
Barristers, 43, 74
Bath, Order of the, 44; classes of, 41; statutes of, 41: *see also* Companions, Dames Commanders, Dames Grand Cross, Knights Commanders, Knights Grand Cross
Bath King of Arms, 79
Beaumont, barony, 109; John, Viscount, 18
Bedford, Jasper of Hatfield, Duke of, 17, 18, 29; John of Lancaster, Duke of, 15
Bennett, Sir Henry, 78-9
Berkeley, barony, 110; William, Viscount, 18
Berners, barony, 104, 111
Bishops, 7, 8, 9, 10-11, 12, 21, 22, 48-52, 72, 120; *in partibus infidelium*, 11; Irish, 11, 50; retired, 52, 120; Welsh, 50; wives of, no ladies: 66: *see also* Durham, Kildare, London, Meath, Suffragan bishops, Winchester
Blackstone, Sir William, 42, 45, 46, 75, 87-8
Blessington, Murrough (Boyle), Viscount, 83
Blood Royal, peers of the, 100: *see also* Dukes of the Blood Royal
Boleyn, Anne, Marchioness of Pembroke, 18
Borthwick, Sir Thomas, 84

132   Index

Botetourt, barony, 104, 109
Botreaux, barony, 104, 110
Bourgh, Sir Thomas, 101
Braga, Council of, 10
Braye, barony, 112
Bristol, aldermen of, 79
British Empire, Order of the, classes of, 41, 68; King of Arms of, 79; statutes of, 41, 69: *see also* Commanders, Dames Commanders, Dames Grand Cross, Knights Commanders, Knights Grand Cross, Members
Bromley, Sir Thomas, Lord Chancellor, 47
Brooke, Ralph, York Herald, 93, 98
Brunswick Lunenburgh, Elector of, 28
Burgh, barony, 112
Burghley, William (Cecil), Lord, Lord High Treasurer, 98
Burke, Sir Bernard, Ulster King of Arms, 89

Caesar, Sir Julius, 59
Cambridge, Adolphus, Duke of, 27, 28-9
Cambridge, mayor of, 77-8
Cambridge University, vice-chancellor of, 77-8
Camden, William, Clarenceux King of Arms, 42-3
Camoys, barony, 107, 110
Canon Law, 4, 20, 23
Canterbury, abp. of, 48, 50, 119; does not pore over the 31st of Henry VIII, 29: *see also* Anselm, Cranmer, Lanfranc
Captains, Royal Navy, 80
Cardinals, 11, 87
Carey, Elizabeth, wife of Sir George, 67
Carr, Sir Robert, 95
Carrington, Charles Robert (Wynn-Carington), Earl, 47
Castlehaven, James (Tuchet), Earl of, 109
Cecil, William, 103
Central Criminal Court, additional judges of, 57
Chancellor of the Duchy of Lancaster, 38, 61, 121
Chancellor of the Exchequer, 38, 61, 121
Chancery, Court of Appeal in, 55
Charles I, 95
Charlotte, Princess, 26
Cheshire Quarter Sessions, 57
Cheyne, Sir John, 101
Chief Baron of the Exchequer, 38, 53
Chief Justice of the Common Pleas, 38, 53

Chief Justice of the King's Bench, 38, 53
Chief Justices, 21
Chivalry, Court of, 15, 45, 90, 91-2; jurisdiction of, 1-2; records of, 92
Cholmondeley, George Henry Hugh, Marquess of, 47
Circuit Judges, 56-7, 70, 73, 122; women, 70-71
Civil Law, 2-3
Clarenceux King of Arms, *see* Camden, Cooke, Lee
Clerk of the Parliaments, 36
Clerk of the Rolls, 19: *see also* Master of the Rolls
Clifford (de Clifford), barony, 103, 104, 108; Henry, Lord (1523), 102-3; Henry, Lord (1627), 103
Clinton, barony, 107, 108; Edward, Lord, 96
Cobham, barony, 109
Cokayne, George Edward, Rouge Dragon Pursuivant, 5
Coke, Sir Edward, 62, 75, 90
Colonels, 45
Commanders of the Order of the British Empire
 men, 122; wives of, 125
 women, 125
Commanders of the Royal Victorian Order, 89
 men, 122; wives of, 125
 women, 125
Committee for Privileges, 36, 50, 96, 97
Common Serjeant, 57, 122
Community Councils, chairmen of, 76-7
Companions of the Distinguished Service Order, 44, 122; wives of, 125
Companions of Honour, 41-2
Companions of the Imperial Service Order,
 men, 122; wives of, 125
 women, 125
Companions of the Order of the Bath, 44, 70, 89
 men, 122; wives of, 125
 women, 125
Companions of the Order of the Indian Empire, 89, 122; wives of, 125
Companions of the Order of St. Michael and St. George, 70, 89
 men, 122; wives of, 125
 women, 125
Companions of the Order of the Star of India, 89, 122; wives of, 125

Comptroller of the Household, 60, 61, 121
Coningsby, Mary, Lady, 81
Constable and Marshal, Court of the, 90, 91: *see also* Chivalry, Court of
Conyers, barony, 111; George William Frederick (Osborne), Lord, 36
Cooke, Robert, Clarenceux King of Arms, 36, 65, 67
Coronations, 14, 15, 16, 58
Corpus Juris Civilis, 2
Countesses, 63, 99, 123
County Councils, chairmen of, 76
County Court Judges, 56, 88
Courts, *see* Appeal, Bankruptcy, Central Criminal Court, Chancery, Chivalry, Constable and Marshal, Crown Court, Exchequer, High Court, Mayor's and City of London Court, Protection, Quarter Sessions, Wards and Liveries
Cowley, Violet, Countess, 64-5
Cranmer, Thomas, abp. of Canterbury, his contempt for suffragan bishops, 51
Cromwell, barony, 110; Thomas, Lord, 24
Crowland Abbey, 86
Crown Court, 56
Cumberland, William, Duke of, 28

Dames Commanders of the Order of the Bath, 70, 124
Dames Commanders of the Order of the British Empire, 125
Dames Commanders of the Order of St. Michael and St. George, 70, 124
Dames Commanders of the Royal Victorian Order, 125
Dames Grand Cross of the Order of the Bath, 69, 124
Dames Grand Cross of the Order of the British Empire, 68, 124
Dames Grand Cross of the Order of St. Michael and St. George, 69-70, 124
Dames Grand Cross of the Royal Victorian Order, 124
Dacre, barony, 110
Danby, Mary, 82
Darcy (de Knayth), barony, 110
Deans, 21, 45
de Clifford, *see* Clifford
De Moleyns, barony, 111
Deputy Lieutenants, 75
de Ros, barony, 103, 104, 105, 106, 107, 112; Robert, 108; William, 112: *see also* Roos

Despenser, barony, 107; Hugh le, 107
Devon (Devonshire), Edward (Courtenay), Earl of, 30; Henry (Courtenay), Earl of, 98; Thomas (Courtenay), Earl of, 96
Dialogus de Scaccario, 4, 8
Disclaimer of peerage, 32, 65
District Councils, chairmen of, 76
Divorced women, 64-5
Doctors, 45; of Divinity, 1, 45
Domesday Book, 8-9, 63
Dorset, John (Beaufort), Marquess of, 15
Dublin, abp. of, 50, 83
Dublin, Robert (de Vere), Marquess of, 15
Duchesses, 98, 123
Dudley, Alice, Duchess, 31; barony, 111
Dufferin and Ava, Maureen Constance, Marchioness of, 64
Dugdale, Sir William, Garter King of Arms, 42
Dukes, 9, 12, 21, 22, 30, 49, 98, 120
 daughters of, 98, 123
 eldest sons of, 98, 120; wives of, 93, 123
 sons of, 15
 younger sons of, 21, 33, 99, 120; wives of, 99, 123
Dukes of the Blood Royal, 20, 29, 100, 119
 daughters of, 123
 eldest sons of, 29, 120; wives of, 123
 younger sons of, 29, 120; wives of, 123
Durham, bp. of, 48, 120
Durham Quarter Sessions, 57

Earl Marshal, 14, 15, 16, 46, 47, 82, 90, 92, 93, 94, 119; Commissioners for executing the office of, 17, 25, 29, 33, 58, 78, 81, 92, 94, 105
Earl Marshal's Books, 81
Earl Marshal's Court, *see* Chivalry, Court of
Earls, 8, 9, 12, 13, 21, 22, 30, 49, 99, 120
 daughters of, 66, 99, 124
 eldest sons of, 33, 99, 120; wives of, 99, 124
 sons of, 15
 younger sons of, 33, 34, 93, 99, 121; wives of, 99, 124
Ecclesiastical law, 4, 23
Ecclesiastical precedence, 4, 20: *see also* Bishops, Suffragan Bishops
Edinburgh, Frederick, Duke of, 28; Prince Philip, Duke of, 27, 119
Edmondson, Joseph, Mowbray Herald Extraordinary, 87

Index 133

134 Index

Edmunds, Sir Thomas, 94
Edward II, 10
Edward IV, Black Book of the Household of, 21; funeral of, 18
Edward the Black Prince, 12
Elizabeth I, 25-6; funeral of, 68
Ellesmere, Thomas (Egerton), Lord, 58
England, baronets of, 39; peers of, 120, 121
Emperor, 87
Erskine, Hon. Thomas, 54
Esquires, 42-5: *see also* Squires
Exchequer, Chancellor of the, 38, 61, 121; Chief Baron of the, 38, 53; Court of, 74; Cursitor Baron of the, 38; postman of the, 74; tub-man of the, 74; Under-Treasurer of the, 38

Fane, Dame Mary, 107
Fauconberg, barony, 111; Thomas (Belasyse), Lord, 31
Fiennes, Richard, 111
FitzGerald, Lady Henry, 107-8
fitz Nigel, Richard, bp. of London, 4, 8
Fitzwalter, barony, 108; Robert (Radcliffe), Lord, 103
Forest Law, 4
Forth, Dorothy, Lady, 65
Furnivall, barony, 108

Garter, Order of the, Chancellor of, 121; Prelate of, 11, 72; statutes of, 72: *see also* Knights
Garter King of Arms, 28, 36, 79, 103: *see also* Dugdale, Heard, Leake, Scott-Gatty, Young
Garter's Roll, 36
Gentlemen, 45; created by the King, 16
George I, 6, 28
George II, 6, 28
George of Cambridge, Prince, 28, 29
George of Denmark, Prince, 26
Glastonbury, abbot of, 12n
Goring, baronetcy, 38
Great Britain, baronets of, 39; peers of, 120, 121
Great Officers of State, 16, 24, 28, 46-8: *see also* Earl Marshal, Lord Chamberlain, Lord Chancellor, Lord Great Chamberlain, Lord High Admiral, Lord High Constable, Lord High Treasurer, Lord President of the Council, Lord Privy Seal, Lord Steward of the Household

Greater London Quarter Sessions, 57
Gresley, Sir George, 38
Grey of Codnor, barony, 108
Grey de Ruthyn, barony, 110

Hale, Sir Matthew, 91
Hanover, George, Electoral Prince of, 28; Sophia, Electress of, 6, 28
Harding-Davies, Mrs G. V., 84
Hastings, barony (1290), 106, 108; (1461), 111; Henry, 105, 106
Hawkins, Serjeant William, 91
Heard, Sir Isaac, Garter King of Arms, 68
Henley, Robert, Lord, Lord Keeper, 53
Henry IV, coronation of, 14
Henry VI, coronation of, 15, 16
Henry VIII, 4, 23, 25
Heralds, 42, 79-80; Extraordinary, 79
High Commissioners, 48, 119
High Court of Justice, 55
  Chancery Division, 56
  Family Division, President of, 55, 121
  judges of, 55-6, 73, 121; women, 71
  Queen's Bench Division, 56
High Sheriffs, *see* Sheriffs
Home Office, 82, 85
House of Lords, journal of, 103-4; seats in, 9, 102: *see also* Clerk of the Parliaments, Committee for Privileges
Howe, Anabella, 82
Hungerford, barony, 111

Indian Empire, Order of the, classes of, 41; statutes of, 41: *see also* Companions, Knights Commanders, Knights Grand Commanders
Inns of Court, 74
Ireland, baronets of, 39; Lord Chancellor of, 36; peers of, 31, 36, 120, 121
James I, 93-4; funeral of, 58
Jennings, ——, J.P. for Plymouth, 1
Judges, 16, 19, 20, 38, 52-8, 121: *see also* Chief Justice(s), Circuit Judges, County Court Judges, Justices, Lords of Appeal in Ordinary, Lords Justices, Master of the Court of Protection, Master of the Rolls, President of the Family Division, Vice-Chancellor(s)
Judicial Precedence, 72-3
Justices, 9, 19, 49: *see also* Chief Justices, Judges
Justices of the Peace, 1, 42, 76

Index 135

Kent, Edmund (Holand), Earl of, 96; Richard (Grey), Earl of, 96
Kent Quarter Sessions, 57
Kildare, bp. of, 50
King, 7, 8, 25; the fountain of honour, 3, 7, 91, 94, 95
  brothers of, 27, 28
  children of, 16
  grandchildren of, 30
  grandsons of, 28
  nephews of, 16, 28
  uncles of, 16, 27, 28
King's Chief Secretary, 60
King's Confessor, 21
Knight-designate, widow of, 83-4, 85, 117
Knights, 18, 40-42, 49
  daughters of, 125
  eldest sons of, 122; wives of, 125
  sons of, 44
  widows of, 65
  wives of, 62, 68
  younger sons of, 122; wives of, 125
Knights Bachelors, 18, 122; eldest sons of, 19; wives of, 125; younger sons of, 19: *see also* Knights
Knights Bannerets, 6, 18, 19; eldest sons of, 19; younger sons of, 19
Knights of the Bath, 19, 41, 95; wives of, 95
Knights Commanders of the Order of the Bath, 41, 70, 122; wives of, 125
Knights Commanders of the Order of the British Empire, 122; wives of, 125
Knights Commanders of the Order of the Indian Empire, 122; wives of, 125
Knights Commanders of the Order of St. Michael and St. George, 122; wives of, 125
Knights Commanders of the Order of the Star of India, 122; wives of, 125
Knights Commanders of the Royal Victorian Order, 122; wives of, 125
Knights Commanders of the Order of St. Michael and St. George, 70
Knights Companions of the Order of the Bath (1725-1815), 41
Knights of the Garter, 19, 38, 40, 72, 121; wives of, 124
Knights Grand Commanders of the Order of the Indian Empire, 122; wives of, 124
Knights Grand Commanders of the Order of the Star of India, 122; wives of, 124
Knights Grand Cross of the Order of the Bath, 41, 69, 121; wives of, 124
Knights Grand Cross of the Order of the British Empire, 68, 122; wives of, 124
Knights Grand Cross of the Order of St. Michael and St. George, 69-70, 122; wives of, 124
Knights Grand Cross of the Royal Victorian Order, 122; wives of, 124
Knights of the Order of the Council, 15-16, 18, 19, 58
Knights of St. Patrick, 121
Knights of the Thistle, 121

Lancashire Quarter Sessions, 57
Lancaster, Chancellor of the Duchy of, 38, 61, 121; Vice-Chancellor of the County Palatine of, 56, 73, 122
Lanfranc, abp. of Canterbury, 10
Laodicea, Augustine, bp. of, 11
Latymer, barony, 102, 111
Law Officers, 19: *see also* Attorney-General, Lord Advocate, Solicitor-General
Leake, John Martin, Garter King of Arms, 67
Lee, Richard, Clarenceux King of Arms, 98
Legate, 21
Leinster, Duchess of, 123
Lennox, Catherine, Duchess, 64
Leopold of Saxe-Coburg, Prince, 26-7
Life Peerage, 3, 33; not an incorporeal hereditament, 33
Life-Peer-designate, widow and children of, 115
Life Peeresses, 124; children of, 67; daughters of, 124; sons of, 35, 121
Life Peers, 121; children of, 67; daughters of, 124; sons of, 35, 121; wives of, 66-7, 124
Liverpool, Recorder of, 56, 122
London, bp. of, 10, 11-12, 48, 120: *see also* fitz Nigel
London, City of, Lord Mayor of, 21, 78; Recorder of, 56, 78, 122: *see also* Common Serjeant, Mayor's and City of London Court
London, Council of (1075), 10; (1176), 10
Lord Advocate, 74
Lord Chamberlain, 16, 19, 46, 47, 119
Lord Chancellor, 16, 19, 20, 21, 27, 46, 47, 50, 52-3, 55, 119: *see also* Bromley
Lord Chancellor of Ireland, 36

Lord Chief Baron, 55
Lord Chief Justice of the Common Pleas, 55
Lord Chief Justice of England, 55, 73, 121
Lord Great Chamberlain, 16, 21, 46, 47, 119; deputy, 47
Lord High Admiral, 16, 46, 47, 119
Lord High Constable, 15, 16, 46, 47, 92, 93, 119
Lord High Treasurer, 16, 19, 20, 27, 46, 47, 119
Lord Keeper of the Great Seal, 53, 119
Lord Lieutenant, 43, 75
Lord President of the Council, 16, 27, 46, 47, 48, 119
Lord Privy Seal, 16, 19, 27, 46, 47, 119
Lord Steward of the Household, 16, 19, 46, 60, 119
Lords of Appeal in Ordinary, 32-3, 55, 121; not peers, 92-3; patents, 32
   children of, 66, 67
   daughters of, 67, 124
   sons of, 34-5; style and title of, 35
   wives of, 66, 67, 124
Lords Commissioners of the Great Seal, 61, 121
Lords Justices, 55, 121
Lords Spiritual, 9
Lords Temporal, 9
Lost grant, presumption of, 4
Lovelace, Richard, Lord, 31

Magistrates, *see* Justices of the Peace
Maids of Honour, 68, 70, 124
Maltravers, barony, 110; Thomas (Fitz-Alan), Lord, 18

Manchester, Recorder of, 56, 122
Manners, Sir Thomas, 102
Marchionesses, 99, 123
Marquesses, 9, 15, 21, 30, 49, 99, 120
   daughters of, 99, 123
   eldest sons of, 15, 99, 120; wives of, 99, 123
   younger sons of, 15, 33, 99, 120; wives of, 124
Married women, 65-6
Mary I, 26
Mary II, 26
Master of the Court of Protection, 57-8, 122
Master of the Horse, 60-61, 119
Master of the Rolls, 38, 53, 55, 73, 121: *see also* Clerk of the Rolls

Masters in Chancery, 57, 58, 88; Extraordinary, 57
Masters in Lunacy, 57, 88
Mayors, 76-7; former, 78: *see also* Reading, Town Mayors
Mayor's and City of London Court, additional judge of, 57
Meath, bp. of, 50
Members of the Order of the British Empire, 44
   men, 122; wives of, 125
   women, 125
Members of Parliament, no precedence as such, 61
Members of the Royal Victorian Order (4th Class)
   men, 122; wives of, 125
   women, 125
Members of the Royal Victorian Order (5th Class)
   men, 69, 122; wives of, 125
   women, 125
Merit, Order of, 42
Michael of Kent, Prince, 29, 119
Middlesex, Rachael, Countess of, 64
Miege, Guy, 62, 65
Milevis, Council of, 10
Milles, Thomas, 43, 87
Milman, Louisa Mary, 110
Moleyns, *see* De Moleyns
Monson, Sir Thomas, 38
Mountjoy, Mountjoy (Blount), Lord, 31
Mowbray, barony, 104, 108; John, Earl Marshal, 14, 96; Roger de, 106; Thomas, Lord, 12
Mowbray Herald Extraordinary, *see* Edmondson

Neville, Richard, 102
Norfolk, John (Mowbray), Duke of, 15; Thomas (Mowbray), Duke of, 96
Norroy King of Arms, *see* Segar
Norroy and Ulster King of Arms, 79
North, Gilbert, 81; Sir John, 81
Nova Scotia, baronets of, 39

Office holders, wives of, 66
Officers of the Order of the British Empire
   men, 122; wives of, 125
   women, 125
Official Referees, 56
'Order of all Estates' (1399), 14-15, 29, 33, 55, 63; ('1479'), 17, 19, 20, 29

Index   137

'Order of all States of Worship' (1429), 15-16, 29
Order of Merit, 42
'Orders according to Ancient Statutes' (1466-7), 16, 50, 53
Orders of Chivalry, *see* Bath, British Empire, Garter, Indian Empire, Royal Victorian Order, St. Michael and St. George, St. Patrick, Star of India, Thistle
'Ordinance or Decree' (1595), 33
Oxford, mayor of, 77-8
Oxford University, chancellor of, 77; vice-chancellor of, 77-8

Parliament, constitution of, 45; seating in, 20, 24
Parish Councils, chairmen of, 76-7
Peer, form of petition for precedence by, 117
Peerage, 30-36; disclaimer of, 32, 65
Peeresses, 65
Peeresses in their own right, sons of, 44, 66
Peers, 18, 30-36
  daughters of, 65, 66
  eldest sons of, 33, 34
  granddaughters of, 67
  grandsons of, 34, 36
  legitimated children of, 35
  younger sons of, 33; daughters of, 125; eldest sons of, 36, 122; wives of, 125
  see also Barons, Dukes, Earls, Life Peers, Marquesses, Viscounts
Pembroke, Anne (Boleyn), Marchioness of, 18
Philip of Spain, Prince, 26
Pope, 25, 87
'Precedence of Great Estates in their owne degre', 17, 63-4
President of the Family Division, 55, 121
Prime Minister, 59, 70, 119
Prince, title of, 29
Prince Consort, 27
Priors, 9, 12, 21, 48
Privy Council, 37, 95
Privy Councillors, 38, 40, 58-9, 121; of Ireland, 92; knighted, 40, 92, 93
Processional, 16
Processions, marshalling of, 5
Protection, Court of, 57-8; Master of the, 57-8, 122
Purchas, John, 77
Pursuivants, 79-80; Extraordinary, 79

Quarter Sessions, chairmen and deputy-chairmen of courts of, 56, 57
Queen, 8, 119, 123; daughter of, 123; sister of, 123; younger sons of, 119
Queen Mother, 123
Queen regnant, 26, 62
Queen's Counsel, 72, 74

Reading, ex-mayors of, 95
Recorders, 76, 78: *see also* Liverpool, London, Manchester
Reeves, 9
Reformation, 5
Rich, Penelope, Lady, 82
Rivers, Anthony (Wydevill), Earl, 17
Roos, barony, 102, 103: *see also* de Ros
Rose, Sir George, 54
Rouge Dragon Pursuivant, *see* Adams
Round, J. H., 46, 112
Royal Family, 7, 25-30
Royal Highness, style of, 29
Royal Household, officers of, 16: *see also* Comptroller, Lord Steward, Master of the Horse, Treasurer, Vice-Chamberlain
Royal Prerogative, 3, 18, 24, 30, 31, 76, 81, 95, 97
Royal Victorian Order, classes of, 41, 44; statutes of, 41, 69; women members of, 69: *see also* Commanders, Dames Commanders, Dames Grand Cross, Knights Commanders, Knights Grand Cross, Members
Royal Warrants, 81-5; fees on, 85; petitions for, drafting of, 82-5; forms of, 113-17
Russell, Edward, Lord, daughter of, 67; John, 21, 52, 61, 86-7

Sadler, Ralph, 60
St. John of Basing, William (Paulet), Lord, 34
St. Michael and St. George, Order of, classes of, 41; King of Arms of, 79; statutes of, 41: *see also* Companions, Dames Commanders, Dames Grand Cross, Knights Commanders, Knights Grand Cross
St. Patrick, Knights of, 121
Saye and Sele, barony, 111
Scotland, baronets of, 39; peers of, 31, 120, 121; precedence in, 6
Scott-Gatty, Sir Alfred, Garter King of Arms, 84

Index

Secretary of State, 16, 24, 60, 120, 121
Segar, William, Norroy King of Arms, 72
Segrave, barony, 108; Nicholas de, 106
Series Ordinum omnium Procerum (1490), 17, 19, 29, 58
Serjeants-at-Arms, 42
Serjeants-at-Law, 38, 45, 49, 58, 88, 93
Seymour, Sir Francis, 81
Shadwell, Sir Lancelot, 54
Sheriffs, 9, 42, 74-5, 76
Sidney, ——, wife of Thomas, 65
Smith, Sir Thomas, 94
Solicitor-General, 73
Somerset, Edward (Seymour), Duke of, 26; John (Beaufort), Earl of, 18; Sarah, Duchess of, 64
Somerset Herald, *see* Treswell
Sophia of Hanover, Electress, 6, 28, 70
Sovereign, *see* King, Queen
Speaker of the House of Commons, 48, 61, 119
Squires, 16
Stafford, Henry, Lord, 96
Stanley, barony, 111
Star of India, Order of the, classes of, 41; statutes of, 41: *see also* Companions, Knights Commanders, Knights Grand Commanders
Stephens, William, 62-3
Stirling-Maxwell, baronetcy, 38
Stonhouse, baronetcy, 38
Strabolgi, barony, 110
Strange, Sir John, 74
Strange of Knokin, barony, 109
Suffragan bishops, 11, 21, 50-52, 120; treated with contempt, 51
Sumner Committee, 107
Supreme Court, 56
Sussex, Augustus Frederick, Duke of, 27

Tables of precedence, 86-9
Talbot, George, 82
Tenants-in-chief, 13, 63
Thistle, Knights of, 121
Thomas, abp. of York, 10
Thoms, William J., 88
Toledo, Council of, 10
Town Mayors, 76-7
Treasurer of the Household, 60, 61, 121
Treswell, Robert, Somerset Herald, 93
Trevor, Sir John, 61
Trinity House, Elder Brethren of, 80
Truro, Thomas (Wilde), Lord, 28

Ulster King of Arms, 36, 42: *see also* Burke
Under-Treasurer of the Exchequer, 38
United Kingdom, baronets of, 39; peers of, 120, 121
Usage, 3-4, 5

Vaux of Harrowden, barony, 107, 112
Vavasour, Sir Charles, 38
Vice-Chamberlain of the Household, 60, 61, 121
Vice-Chancellor, vii-viii, 56
Vice-Chancellor of the County Palatine of Lancaster, 56, 73, 122
Vice-Chancellor of England, 53
Vice-Chancellors, 54
Vicegerent in spirituals, 24, 119
Victoria, Queen, 27, 28; exceeded her power, 3
Viscountesses, 123-4
Viscounts, 9, 17-18, 21, 22, 30, 33, 49, 99, 120
daughters of, 100, 124
eldest sons of, 34, 99, 121; wives of, 99, 124
sons of, 33
younger sons of, 33, 37, 38, 100, 121; wives of, 100, 124

Wales, English law applied to, vii; precedence in, 76
Wales, Prince of, 119; eldest son of, 30
Wales, Princess of, vii
Walpole, Mary, 83
Wards and Liveries, Master of the Court of, 6, 38
Warwick, Richard (Beauchamp), Earl of, 14, 96
Webb, Dr William, 77
Wellington, Arthur (Wellesley), Duke of, 27
Wentworth, barony, 112
Westminster, abbot of, 12n
Westmorland, Ralph (Neville), Earl of, 98
Widows, 64-5
William I, 7
William, III, 26
William of Orange, Prince, 26
Willoughby, Richard, 102
Willoughby de Broke, barony, 107, 111
Willoughby de Eresby, barony, 109
Winchester, bp. of, 10, 11-12, 48, 120
Windsor Herald, *see* Ashmole

Worcester, Charles (Somerset), Earl of, 17, 98; John (Tiptoft), Earl of, 50, 53
Wriothesley, Thomas, 60
Women, precedence of, 62-71
Writ of acceleration, 34
Writ of summons, 101

Yeomans, Sir Robert, 95

York, abp. of, 10, 48, 50, 119
York, Ernest Augustus, Duke of, 28
York Herald, *see* Brooke
Young, Sir Charles, Garter King of Arms, 44, 75, 88

Zouch(e) of Harringworth, barony, 104, 109